Friederich Kupzog

Frequency-Responsive Load Management in Electric Power Grids

Friederich Kupzog

Frequency-Responsive Load Management in Electric Power Grids

A Technical Concept for Demand Response in Smart Grids

Südwestdeutscher Verlag für Hochschulschriften

Impressum/Imprint (nur für Deutschland/ only for Germany)
Bibliografische Information der Deutschen Nationalbibliothek: Die Deutsche Nationalbibliothek verzeichnet diese Publikation in der Deutschen Nationalbibliografie; detaillierte bibliografische Daten sind im Internet über http://dnb.d-nb.de abrufbar.
Alle in diesem Buch genannten Marken und Produktnamen unterliegen warenzeichen-, marken- oder patentrechtlichem Schutz bzw. sind Warenzeichen oder eingetragene Warenzeichen der jeweiligen Inhaber. Die Wiedergabe von Marken, Produktnamen, Gebrauchsnamen, Handelsnamen, Warenbezeichnungen u.s.w. in diesem Werk berechtigt auch ohne besondere Kennzeichnung nicht zu der Annahme, dass solche Namen im Sinne der Warenzeichen- und Markenschutzgesetzgebung als frei zu betrachten wären und daher von jedermann benutzt werden dürften.

Verlag: Südwestdeutscher Verlag für Hochschulschriften Aktiengesellschaft & Co. KG
Dudweiler Landstr. 99, 66123 Saarbrücken, Deutschland
Telefon +49 681 37 20 271-1, Telefax +49 681 37 20 271-0, Email: info@svh-verlag.de
Zugl.: Wien, TU, Diss., 2008

Herstellung in Deutschland:
Schaltungsdienst Lange o.H.G., Zehrensdorfer Str. 11, D-12277 Berlin
Books on Demand GmbH, Gutenbergring 53, D-22848 Norderstedt
Reha GmbH, Dudweiler Landstr. 99, D- 66123 Saarbrücken
ISBN: 978-3-8381-0125-5

Imprint (only for USA, GB)
Bibliographic information published by the Deutsche Nationalbibliothek: The Deutsche Nationalbibliothek lists this publication in the Deutsche Nationalbibliografie; detailed bibliographic data are available in the Internet at http://dnb.d-nb.de.
Any brand names and product names mentioned in this book are subject to trademark, brand or patent protection and are trademarks or registered trademarks of their respective holders. The use of brand names, product names, common names, trade names, product descriptions etc. even without
a particular marking in this works is in no way to be construed to mean that such names may be regarded as unrestricted in respect of trademark and brand protection legislation and could thus be used by anyone.

Publisher:
Südwestdeutscher Verlag für Hochschulschriften Aktiengesellschaft & Co. KG
Dudweiler Landstr. 99, 66123 Saarbrücken, Germany
Phone +49 681 37 20 271-1, Fax +49 681 37 20 271-0, Email: info@svh-verlag.de

Copyright © 2008 Südwestdeutscher Verlag für Hochschulschriften Aktiengesellschaft & Co. KG and licensors
All rights reserved. Saarbrücken 2008

Produced in USA and UK by:
Lightning Source Inc., 1246 Heil Quaker Blvd., La Vergne, TN 37086, USA
Lightning Source UK Ltd., Chapter House, Pitfield, Kiln Farm, Milton Keynes, MK11 3LW, GB
BookSurge, 7290 B. Investment Drive, North Charleston, SC 29418, USA
ISBN: 978-3-8381-0125-5

Kurzfassung

Im elektrischen Energieversorgungsnetz, das sich von Generatoren zu elektrischen Lasten erstreckt, muss zu jedem Zeitpunkt die exakte Balance von Erzeugung und Verbrauch eingehalten werden. Die Liberalisierung der Energiemärkte und die zunehmende Nutzung erneuerbarer Energieträger führen zu einem steigenden Bedarf an Regelenergie aufgrund der dynamischen Erzeugerprofile verteilter Erzeugung. Zurzeit wird die Balance hauptsächlich durch Maßnahmen auf der Erzeugerseite gehalten. Energie kann aber auch auf der Verbraucherseite vorgehalten werden, entweder nur konzeptionell in Form von Lastverschiebungen oder tatsächlich, z. B. in Form von thermischer Energie bei Heiz- oder Kühlprozessen. Koordiniertes Energie-Management auf der Verbraucherseite des elektrischen Netzes hat das Potential, die konventionellen Methoden der Balanceregelung zu revolutionieren. In dieser Arbeit wird ein neues Modell für das Verhalten solcher verteilter Energiespeicherung in Lastprozessen abgeleitet. Das Modell beruht auf Fallstudien und Messungen von elektrischen Verbrauchern, bei denen träge thermische Prozesse ablaufen. Basierend auf diesem Modell wird ein verteilter Algorithmus entwickelt, der es ermöglicht, primäre Balanceregelung im Netz durch Lastmanagement zu übernehmen. Dieses System wird in einer Simulationsumgebung implementiert, getestet und mit verwandten Lösungen verglichen. Die technische Umsetzbarkeit von lastseitiger Primärregelung wird dabei demonstriert. Der vorgeschlagene Algorithmus ist robust gegen Ausfälle der Kommunikationsinfrastruktur, da er in erster Linie auf Schwankungen der Netzfrequenz reagiert, und die Kommunikation zwischen den Lastknoten keinen Echtzeitbedingungen genügen muss. Er erlaubt Energieverbrauchern, Primärregelenergie für das elektrische Netz bereitzustellen, ohne dadurch Komforteinbußen in Kauf nehmen zu müssen. Dadurch ermöglicht er es Energiekunden, finanzielle Vorteile aus den Erlösen für Primärregelenergie zu lukrieren, sollte der Marktzutritt für Endverbraucher geöffnet werden. Das vorgeschlagenen System ist ein Schritt in die Richtung zukünftiger Energiesysteme, die mit beträchtlich mehr Informationstechnologie ausgestattet sein werden, als es heute der Fall ist, um die kommenden Herausforderungen elektrischer Energieversorgung zu erfüllen.

Abstract

In the electric power grid, which stretches from generators to electric loads, the exact balance between supply and demand has to be maintained at all times. The liberalisation of power markets and the rising utilisation of renewable energy sources have lead to a growing demand for imbalance energy due to the dynamic patterns of distributed generation. Today, the energy balance is only maintained by measures on the generation side. However, energy can also be stored in power consuming processes on the demand side, either conceptually by load shifting, or in inert (e.g. thermal) consumption processes. Coordinated energy management measures on the demand side have the potential to revolutionise the techniques of maintaining the power balance in the grid. This work proposes a new approach for modelling the properties of such distributed energy storing processes. The model is supported by case studies and measurements of loads incorporating inert thermal processes. Based on the model, a distributed algorithm for primary control on the basis of demand response is proposed, implemented and compared to related approaches. Thereby, the technical feasibility of demand-side balance control, which is a key application in future energy systems, is shown. The algorithm for frequency-controlled demand side management deployed in this work is robust against communication faults and synchronisation errors, since it primarily reacts on changes of the ubiquitous grid frequency, and the communication between the managed loads therefore does not need to be in real-time. It enables energy consumers to commit to primary control without noticeable loss of comfort in their electricity availability. Further it permits energy customers to gain financial profits from the compensations paid for imbalance energy provision, if the market access is opened for energy end users. The proposed system is one step on the way to future energy systems that will be equipped with considerably more information technology than it is the case today in order to meet the upcoming challenges of electrical energy supply.

Preface

The research for this thesis has astonished me at several stages with positive surprises. It began with the chance to move to Vienna. Coming from a computer technology background, I became fascinated by the fundamental mechanisms of the electric power grid. I saw how the basic principles and techniques of distributed systems have been applied in its design long before these principles were found and written down for computer systems. Later in the course of my work, larger obstacles disappeared. I was able to avoid real-time communication in my control algorithm by using the grid frequency for this purpose. Two modelling problems suddenly cancelled out each other. All these events helped me to keep on the track and to go the next constructive step. Therefore, I wish the reader that such positive surprises also occur in her or his work.

Friederich Kupzog, Vienna, June 2008

Table of contents

1. Introduction 11
   1.1 Motivations for change in electrical energy supply 11
   1.2 Balancing the power grid 12
   1.3 New role for the demand side 14
   1.4 Scope, challenge and task of this work 15

2. Technical background 17
   2.1 Generators and frequency control 17
      2.1.1 Generator with isochronous governor and with droop characteristic 19
      2.1.2 Multiple generators and loads 21
      2.1.3 Generator for secondary response 22
   2.2 Electric loads 23
   2.3 Differential equations for time-domain simulations 25
      2.3.1 Inertia 25
      2.3.2 Primary and secondary response 26
      2.3.3 Example of sudden increase in load: simulation model and results 27

3. Related work 31
   3.1 Previous research in the context of this thesis 31
   3.2 Energy storage for neutralising load fluctuations 35
   3.3 Frequency-responsive loads 36
      3.3.1 "GridFriendly^{TM} Appliance Controller": threshold-based algorithm 37
      3.3.2 Approaches of continuous load modulation on grid-frequency basis 38
   3.4 Classification of demand side management 40
   3.5 Load shifting potential in Austria 41

4. Modelling resources for demand side management 45
   4.1 Modelling approach using conceptual storages 45
      4.1.1 First approach for a generic model of load shifting 47
      4.1.2 Concept of energy packets resulting in two dimensions of freedom 50
   4.2 Case study: following required load profile 53
      4.2.1 Optimal scheduling of distributed storages 53
      4.2.2 Formulation of the linear optimisation problem 54
      4.2.3 Solving the optimisation problem 57
      4.2.4 Solution to an example problem 58
   4.3 Model refinement for inert thermal processes 60
      4.3.1 Equivalent electrical model for a thermal process 60
      4.3.2 Discussion of thermal and electrical losses 62
      4.3.3 Model verification by measurements 63
      4.3.4 Two options for active load shifts 65
   4.4 Model comparison using case study 68

4.5 Distributed thermal resource model ... 72
5. Primary control algorithm using distributed resources ... 75
   5.1 Deduction of the algorithm ... 75
      5.1.1 Discussion of the simplified approach with $b_i = 0$ ... 78
      5.1.2 Structure of the resulting system ... 80
      5.1.3 Improved algorithm for the case $b_i > 0$ ... 82
      5.1.4 Time-discrete realisation ... 86
   5.2 Grid frequency ... 87
   5.3 Workload balancing by activation level update ... 88
      5.3.1 Simulation and comparison of permutation algorithms ... 90
      5.3.2 Redundant server system ... 93

6. Simulation results and comparisons ... 97
   6.1 Statistical analysis of performance simulations ... 97
      6.1.1 Simulation setup and trajectory visualisation ... 97
      6.1.2 Quality measure for performance measurement ... 100
      6.1.3 Optimisation of τ ... 103
      6.1.4 Slow resource reaction ... 104
      6.1.5 Influence of storage discharges ... 106
   6.2 Dynamic analysis ... 108
      6.2.1 Simulation scenarios and underlying models ... 109
      6.2.2 Result of dynamic simulation ... 111
      6.2.3 Improvement of the dynamic behaviour ... 112
      6.2.4 Comparison with the approach of Short et al. ... 114
      6.2.5 Comparison with the GridFriendly Appliance Controller ... 115
      6.2.6 Discussion ... 117

7. Implementation ... 119
   7.1 System design considerations ... 119
   7.2 DSM interface unit ... 121
   7.3 Central server implementation and communication protocol ... 127
   7.4 Implementation results ... 128

8. Conclusions and Outlook ... 131
   8.1 Main contribution ... 131
   8.2 From demand side management to control power provision ... 132
   8.3 Defining factors for the DSM algorithm performance ... 133
   8.4 Steps towards a realisation of control power from the demand side ... 135
   8.5 Information technology in the power grid: the paradigm shift ... 137
   8.6 Vision of smart electricity grids ... 139

Appendix A: Resource sets ... 143

Appendix B: DSM interface unit schematics ... 146

Appendix C: Communication protocol ... 148

References on scientific publications ... 161

Internet references ... 165

Abbreviations

CHP	Combined Heat and Power
CPU	Central Processing Unit
DG	Distributed Generation
DIN	Deutsche Industrie Norm
DSM	Demand Side Management
EEPROM	Electronically Erasable Programmable Read Only Memory
FPGA	Field Programmable Gate Array
GFA	Grid-friendly Appliance
GPRS	General Packet Radio Service
GSM	Global System for Mobile Communications
HSDPA	High Speed Downlink Packet Access
IEEE	Institute of Electrical and Electronics Engineers
IRON	Integral Resource Optimisation Network
IRON-CPP	Integral Resource Optimisation Network – Control Power Provision
IRON-CPP-PI	Integral Resource Optimisation Network – Control Power Provision – Proportional/Integral
ITU	International Telecommunication Union
KNIVES	Keio University Network oriented Intelligent and versatile Energy saving System
OECD	Organisation for Economic Co-operation and Development
PLC	Power Line Communication
PLL	Phased Locked Loop
RAM	Random Access Memory
SCADA	Supervisory Control And Data Acquisition
SCIP	Solving Constraint Integer Programs
SOAP	Simple Object Access Protocol
SSID	Service Set Identifier
UCTE	Union for the Co-ordination of Transmission of Electricity
WLAN	Wireless Local Area Network
XML	Extensible Markup Language
ZIMPL	Zuse Institute Mathematical Programming Language

1. Introduction

Electrical energy has become a commodity of modern life, such that in some parts of the world its availability is not even noticed anymore. It is also a central and basic requirement for nearly all economic activities. The exploitation of fossil energy resources is still growing. The finiteness of fossil energy resources is currently countered by a steady improvement of technologies for finding new repositories and accessing fields more and more complicated to exploit. This holds for resources that are used for producing electricity as well as for all fossil primary energy resources in general. Although these technological advances shift the point in time further into the future when resources will be technically exhausted, this development is working towards an inescapable end for the simple reason that the exploitation of natural fossil resources happens many orders of magnitude faster than these resources have been developing. However, the exploitation will continue for some time due to the lack of sufficient alternatives available today.

1.1 Motivations for change in electrical energy supply

While the described development has now been foreseeable for many decades, a new aspect has gained attention more recently, namely the insight that the emission of the so-called green house gases (specified in [Kyoto98, p. 19]), which are set free in energy conversion process based on fossil resources, has an impact on the global climate. In 1997, this understanding was formulated for the first time on a broad basis in the Kyoto Protocol [Kyoto98], which was signed by representatives of a substantial number of nations. Since then, many national and international targets for cuts in green house gas emissions have been set, defining requirements for reduction of emissions over the coming decades.

These two main factors – the finiteness of fossil energy resources and the effects of emitted green house gases – are strong motivations for changes in the way of gaining and using energy. As new considerable and fast-growing energy consumers join the old established ones (as Asia and India currently do), a third issue occurs: the exploitation rate. It is determined on the basis of political decisions and not by technical considerations. This fact results in the emergence of a third motivation for changes in the sources and use of energy: independence of supply and energy autarky are again considered as important strategic objectives.

Introduction

In many countries, including the member states of the European Union [1, 2], Japan [3] and parts of the USA [4], it can be observed that governments consider it currently as their task to stimulate the technical advance and innovation in the context of electricity generation and use, resulting in a wide range of research activities in this area. It should be noted that most research currently carried out in the area of energy systems is the result of strategic political decisions.

Basically, three different strategies are currently discussed that are able to tackle the upcoming energy challenge. One of these strategies is behaviour-oriented and intends to influence the end-consumers to use less energy. This is not primarily a technology-oriented strategy but relies on social and behavioural factors. However, technology can increase the motivation to save energy, e.g. by relaying real-time energy prices on-line to the end consumer. The remaining two approaches are clearly technology-oriented and refer to the way improved and new technologies can help to reduce the use of (and finally replace) electricity gained from fossil resources. These are efficiency measures and alternative energy sources. Efficiency measures refer to the fact that more efficient end-user equipment (such as better isolated refrigerators or homes, energy-efficient pumps etc.) and generation technologies (such as combined heat and power, CHP) help to reduce the amount of primary energy needed. In terms of alternative energy sources, technologies considered are electricity generation from renewable energy resources such as wind, water, biomass, solar, tidal power etc., nuclear energy and in future maybe fusion energy. Still, the future composition of this portfolio is subject to strong discussions. The sustainability of some of these options is still questionable (e.g. biomass). The problem of nuclear waste and the danger associated with atomic reactors leads to different attitudes towards nuclear energy, changing over time and place. Nevertheless, it is a widely accepted understanding that renewable energy resources will play a substantial role in future, having a considerably higher share of the total energy exchange than they have today.

1.2 Balancing the power grid

Compared to the traditional generation from fossil resources, the energy density of renewable energy sources is low. The number of generation units is comparably high, but they have a rather low individual power output compared to large centralised power plants. Units are set up at locations where the availability of the energy source is good (e.g. strong winds, flowing water). This results in generation units being scattered over the power grid infrastructure in a spatially distributed manner. Therefore, this kind of electricity generation is often called "Distributed Generation", or simply referred to as DG.

The integration of a high density of distributed generation into the existing power grids leads to a number of different issues, which are currently discussed internationally amongst distribution grid operators, researchers, plant operators, technology providers and grid regulation authorities [Lug07]. One of these issues is the fact that a strong growth of electricity generation in the medium voltage grid, where most of the installed distributed generation injects its power, leads to grid voltage problems [Jen00]. In times of low demand, the grid voltages at the feeding points reach the limits set by grid operators and regulation authorities, so that no more units could be installed without significant grid investments [Kup07a].

A second issue is that due to the volatile nature of generation from renewable energy resources, it becomes more and more difficult to predict the amount of electricity generation with a growing share of distributed generators in the power system. As a result, in future there will be a stronger need for balancing energy in the grid as there is today [Str02]. This is the central motivation for this work. The balance of electricity generation on one side and electricity consumption on the other side is today maintained by control actions on the generation side. It is the task of the large generators to adapt to the changing load situation and always feed exactly as much energy into the system as is needed. In practice, power plants are scheduled on the basis of a day-to-day forecast for electricity demand and the difference between predicted and actual power consumption is outweighed by dynamic control actions. Distributed generators, especially wind generators, do not contribute to these control actions but feed energy into the grid when it is available. Despite good wind power predictions, the rising amount of distributed generation will significantly increase the uncertainty in balance prediction and therefore increase the need for more balance energy provision [Short07].

Since the balance control will become more and more dynamic due to the strongly varying generation patterns of generation from renewables, there is an increasing need for energy storage in the grid [Stad03a], which also results in increasing costs for balance energy [Str02]. According to Black, Silva and Strbac, the costs for balancing and balancing capacity dominate the costs of transmission and distribution of additional distributed generation [Bla05]. Energy storage is an attractive solution to the balancing problem since it enables generators to feed the grid whenever this is technically or economically attractive, without carrying the burden of maintaining the exact accordance of generation and demand at all times. In Europe, existing hydro storages cannot satisfy all of this expected growing demand since most are working to full capacity fulfilling currently existing needs [Rup07]. The introduction of a "super grid" that connects grids in different time zones and climatic zones and thus improves the fluctuation of electricity over long distances would reduce the need for storage; however building such a super grid infrastructure as well as setting up new hydro powers plants is associated with significant investments and considerable administrative challenges [Stad03a]. Nevertheless, the technical need and the rising revenues associated with balance energy provision result in a number of new hydro storages. An example is Kops II, a 450 MW hydro storage in western Austria that primarily is build to satisfy the balance energy needs of German wind power, supposed to go into full operation in 2008, with investments costs of about 370 Million Euro [5].

In opposition to large-scale approaches such as super grids and large hydro storages, there are also concepts of tackling the balance issue of distributed generation with distributed storages. Local storages that are situated closely to the generator allow for more flexibility in feed-in times for power from renewables. Storage technologies considered are batteries (e.g. Vanadium Redox batteries

Introduction

[Sch05]), flywheels ([Heb02]), super capacitors, superconducting magnetic energy storage, or even hydrogen generation [Sta03]. Setting up this distributed solution has less administrative challenges than large centralised hydro storages, but the economic efficiency is lower.

1.3 New role for the demand side

In this situation of growing need and growing compensations for balance energy provision, it is worth to consider the potential contribution of the demand side in the power system. Here, large unutilised potentials for energy storage (e.g. in inert thermal processes for refrigeration, air conditioning and heating) exist, that could help to maintain the power balance in a control area [Sta03, Inf07]. These demand side processes convert electrical energy in some other form of energy (thermal energy, chemical energy, potential energy, rotation energy etc.) and store the energy in this form, e.g. in a thermally isolated room. Demand side storages cannot feed back energy into the grid, but they can store the energy needed by the end-user process so that the timing of energy consumption becomes more flexible. By extending control strategies to the demand side, more degrees of freedom are made available in the overall system, resulting in a more effective and efficient operation. So far unused potentials are utilised. The great advantage of using storages in demand side processes is that these storages are sometimes already existing and do not have to be newly erected. The only additional costs of associated with the use of these demand side storages is that if the communication infrastructure needed to control and coordinate the operation of these storages (basically the timing of charge and discharge activities). Here, many smaller investments are necessary compared to one huge investment needed for realising a large hydro storage. Further, a distributed solution can be more reliable because there is no single point of failure [Kirby03, p. 17], and transmission losses are reduced in comparison to a centralised solution. All this makes the option of distributed demand-side storages advantageous and therefore attractive.

For balance energy provision on the demand side of the energy system, energy consumption of electrical loads is not reduced in terms of energy savings. The consumption patterns are rather changed in such a way that the interaction with the power grid is optimised. This results in short-term modulation of the original, non-influenced consumption patterns, so that demand is increased in certain time intervals and decreased in others. The mere short-time shift of electricity demand enables to replace balance energy provision by gas power plants. CO_2 emissions caused by set-point changes in these power plants due to balance control actions can be avoided [Bra06 p. 61].

The potential of loads in the system that can be influenced and used as distributed demand-side storages is sufficiently large. In Germany, electricity consumption for cooling (industrial and domestic) is about 66 TWh/a and already accounts for nearly 10% of the total electricity consumption [Sta03]. Other countries having a much higher density of air conditioning systems reach even higher percentages. The utilisation of some of the existing demand side potentials such as electrical space or water heating is an established technique since the 1960s [Short07] to shift demand from peak to off-peak times. However, in contrast to this coarse-grain approach, where only large groups of loads can be switched either remotely by radio or power line signals or timer-based at fixed times, a solution for balance energy provision has to be much more flexible, must be able to quickly adapt to changing situations in the grid and operate completely autonomously and automatically, without any need for manual intervention by the energy end-user or grid operator.

It is important that the measures of influencing the energy consumption of the end-user load are completely hidden from the end-user. Only then the technology will be accepted and has a chance to be widely used [Kup06]. For the load management operation to be hidden from the user, the time constants of the load management must be shorter than the time constants of the user process. By this it is guaranteed that load management measures do not interfere too strongly with the user process.

1.4 Scope, challenge and task of this work

The scope of this work is the utilisation of energy-storage capabilities in electricity-consuming processes for the sake of providing these storage capabilities to the power grid, which gains flexibilities in the timing of power consumption from this. These flexibilities can be used to increase the efficiency of the overall energy system and eases the integration of generation from renewable energy sources. The means of utilisation is computer and communication technology.

The specific challenge of this approach lies in the individuality of power-consuming processes and their flexibilities. Before any clear and concise scientific assessment of the control technology needed for utilising the load shifting potential which is available on the demand side, it seems to be of high complexity and having to deal with an nearly endless list of process classes, each of which has numerous parameters describing the complex behaviour of the individual electrical load and the operational boundary conditions of the consumption process. This work does not deal with the assessment of the load shifting potential, since this challenge has already been taken up in previous works, e.g. [Stad03a] or [Bra06] and is not related to computer technology. It rather starts after this point with the problem of designing the technology needed to utilise the existing potentials. The first sub-challenge in this is to find a generic way of modelling the behaviour of many different individual consumption processes with load shift potential. The second sub-challenge is to design a robust, inexpensive and scalable automation infrastructure that serves as basis for executing algorithms that schedule the load shifts in such a way that they are of advantage for the power grid operation.

Previous works [Pal06, Kup08] have shown that the most attractive application of demand-side energy storage lies (in Austria and Europe, today) in the provision of imbalance energy (see chapters 2 and 3). The task of this work is therefore to find a feasible technical solution that enables the participation of electrical loads in balance energy provision and to verify the solution in simulations. The resulting system has to comply with the European rules of primary control power provision

[UCTE04a, UCTE04b] and furthermore operate in such a way that the interference with user processes constrained to an acceptable minimum. Therefore, the focus shall be put on inert demand-side processes with energy storage capabilities. In order to gain considerable amounts of balance energy, a high number of individual processes have to be included in the system.

The main questions motivating this work therefore are:

- *How can primary control power be provided by energy consumers?*
- *How can the control algorithm look like?*
- *What infrastructure is necessary for that?*
- *How complex and costly is the hardware needed at the individual electrical loads?*

In the following, these questions will be investigated and answered.

2. Technical background

This work proposes a new scheme for the active contribution of the power system's demand side to the on-line balancing of supply and demand. Therefore, in the following the conventional principles of balance-keeping in the power grid are reviewed as a basis for subsequent discussions. Further, also the fundamental technical concepts that enable the distributed operation of generators and loads in the power grid are reviewed. At the same time, the mathematical models for power system resources which are used throughout this work are introduced. The power-engineering basics presented in the following are an excerpt from [Kund94], if not otherwise indicated.

2.1 Generators and frequency control

Today's large interconnected power grids have evolved step by step from smaller, local utilities that originally supplied only local communities. In the history of power systems it became clear quite early that by interconnecting the isolated systems the economic efficiency and reliability of power supply is strongly improved [Casaz04]. The obvious engineering challenge of the electrical power grid lies in the creation of electrical structures in such a scale that they can span over countries and even continents. However, from an automation and control point of view, the primary challenge of the power system lies not in its size itself but in its detailed design that allows a stable and reliable operation of such a large and complex system.

The electric power grid is a classical distributed system [Tan06, p. 2], even if it is not consisting of computers but of electrical generators and loads. Seen from an abstract viewpoint, it consists of a large number of interacting entities (nodes or energy resources) that are connected with communication channels, the power lines. Communication is realised by influencing and observing ubiquitous physical parameters like power flows or the grid frequency. The growing power grid was, together with the telephony system that had its growing period nearly at the same time [Casaz04, Casso04], one of the first and largest distributed systems that have been designed by electrical engineers. This duality of key infrastructure systems, one for electrical energy and the other for communication, which began in the 1880s and 1890s, still exists today. However, while the telephone system has merged into the Internet and thus made through a number of revolutionary technological changes, the power system has remained with comparably little changes. The primary reason for this is that changes in the power grid are associated with extraordinary high investment costs (compared to those of smaller components in the telecom sector), resulting in long investment cycles of several decades. Consequently, the technology is designed to stay in place and operate for many years. Thus, the technical solutions used in the grid tend to be extremely well-considered in the first place, but due to the low replacement rate technical innovation take a long time to be put in practice.

The electric power grid cannot store electrical energy in large amounts for long times. This basic fact has wide-reaching effects on the design of the grid. The generated power has to match the consumed power at all times. For the power grid seen as a distributed system, this requirement results in the need for real-time communication between the generators and the loads in the system. In some way, the current demand has to be communicated to the generation sites, so that they can adjust to it (theoretically, also the generation amount could be communicated to the loads so that they would adjust to it). Further, a suitable way of load-sharing between the generators is needed, which can be seen as a protocol that determines in detail which generator reacts when and how much. All this has to work over hundreds or even thousands of kilometres. This technical challenge was brilliantly met without the use of any explicit data communication by the invention of power-frequency control.

In order to illustrate the principle of power-frequency control, a simple setup with a single electric generator and a load shall be considered first. In this example, the generated power shall origin from a rotating turbine. The vast majority of electricity comes from turbines connected to generators. Turbines are rotating structures that are accelerated by a flow of gas or liquid. The connected generator generates alternating current on the basis of the dynamo-electric effect. The shaft between turbine and generator rotates with a specific rotation speed ω_R that is determined by the pressure on the turbine on one side (accelerating) and the electric load of the generator on the other side (decelerating). Additionally, the mechanically rotating parts (masses) have the inertia I. The energy stored in this rotation is

$$E_{rot} = 0.5 \, I \, \omega_R^2. \tag{1}$$

Consequently, even if the pressure on the turbine suddenly drops, the shaft will continue to rotate for several seconds or even minutes. These rotating masses of the generators are the only way energy can be stored in the power grid. Factually the storage happens on the generator site and not in the grid, which strictly speaking consists only of power lines and transformers, but for convenience the generator inertia is counted as a feature of the grid here.

The block diagram in Figure 2.1 shows the above-described system in a control-engineering fashion. $H_T(s)$ is the transfer function of the turbine and connected generator, $H_I(s)$ is that of the inertia of the rotating masses (see Section 2.3.1). The gas or liquid pressure on the turbine is determined by the valve setting v_{set} and changed into electric power P_{gen}. The electric load P_{load} is subtracted from this. If there is any difference remaining, this causes a change in rotation speed. The rate of this change is restricted by the mechanical inertia.

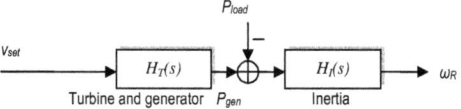

Figure 2.1: Open-loop model of a power generator

Obviously, the rotation speed ω_R is nothing else as the frequency of the alternating current generated. For numerous reasons, it is not desirable that the grid frequency changes considerably all the time. This is because most appliances and transformers are optimised for a certain nominal frequency. Further, the rotation speed of electrical machines is determined by the frequency and these machines should have a constant rotation speed. Also, simple electronic clocks derive the time from the oscillation of the grid voltage [UCTE04a]. For all these reasons, the grid frequency should not deviate too far from the nominal value and should in average be very close to the nominal value. Therefore, instead of being an open-loop system as shown in Figure 2.1, the generator is rather embedded into a control loop that controls the generated power in such a way that a certain grid frequency ω_{nom} is maintained ($f_{nom} = \omega_{nom}/2\pi = 60$ Hz in the US and 50 Hz in Europe and Asia).

2.1.1 Generator with isochronous governor and with droop characteristic

For maintaining the grid frequency, a simple isochronous governor could be used (see Figure 2.2). This integral controller k/s integrates the frequency error and controls the turbine pressure accordingly so that no steady-state frequency deviation occurs.

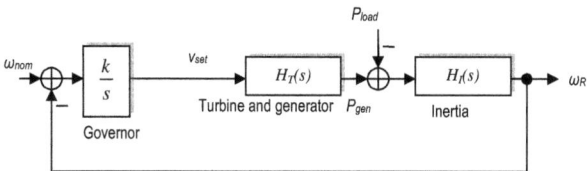

Figure 2.2: Generator with isochronous governor

This approach is practically applied in backup-power generators. It can be used, when only a single generator is supplying the load. The disadvantage of this method of frequency stabilisation is that as soon as multiple generators are connected to the same grid, the controllers interfere with each other and even can work against each other, so that no stable frequency is reached. The controller has to be designed differently in order to allow a distributed operation.

Technical Background

In contrast to the isochronous governor design, the actual controllers for frequency stabilisation used in the power grid allow for a small steady-state frequency-deviation in the range of several mHz for the sake of stability. The generated power P_{gen} is increased when the frequency falls and decreased when the frequency rises according to the so-called "droop" characteristic shown in Figure 2.3. As a consequence, the grid frequency is no longer controlled directly, but only the generated power of each unit connected to the grid is controlled *according* to the grid frequency. Therefore, this technique is called *power-frequency control*. The approach of power-frequency control allows joint frequency stabilisation with multiple generators that can be several hundred kilometres apart from each other.

The steady-state output power P_{tar} of a generation unit can be written in the form

$$P_{tar} = a(f - f_{nom}) + b, \qquad (2)$$

(compare Figure 2.3) where a indicates the slope of the droop characteristic and $b = P_{nom}$ indicates the generation power at the nominal frequency (50 Hz or 60 Hz). A typical droop is dimensioned in such a way that for a 4 % frequency increase the generated power doubles [Short07]. For a $P_{nom} = 100$ MW power station this results in $a = 50$ MW/Hz.

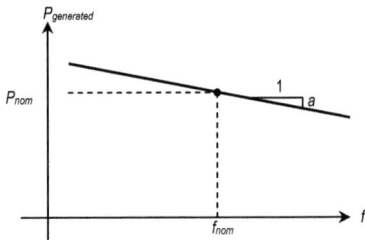

Figure 2.3: The generated power changes with the grid frequency

In (2) however only the steady-state output power is described. The dynamic reaction on load and frequency changes is modelled by the system of transfer functions shown in Figure 2.4 and Figure 2.5.

The governor is constructed as shown in Figure 2.4 with a proportional feedback constant R, which causes the linear droop. ($R = [-2\pi a]^{-1}$, this is shown in Section 2.3.2). The feedback loop can be replaced by an equivalent serial system (see Figure 2.4 b and c). The simple isochronous governor from Figure 2.2 is replaced by the resulting droop governor as shown in Figure 2.5. Now, the system input is the nominal power P_{nom} instead of the nominal frequency f_{nom}. The nominal power is the power generated at nominal frequency under steady-state conditions. In the following, nominal conditions are assumed to be the normal working point of the system. Therefore, the control variable is no longer ω_R, but $\Delta\omega_R = \omega_R - \omega_{nom} = 2\pi(f - f_{nom})$.

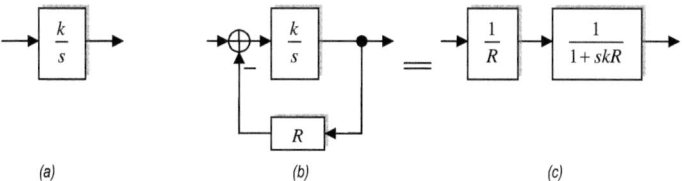

Figure 2.4: The droop characteristic is realised by adding a proportional feedback to the isochronous governor. (a) Isochronous governor (b) Governor with proportional feedback (c) Equivalent system without feedback connection

Generators equipped with a droop governor are reacting fast on frequency changes in the grid (within several seconds) Therefore, their reaction is called "primary response" or "primary control" or "frequency response". Primary response is the first measure that becomes active as soon the grid frequency drops or rises due to a significant change in load or generation. It retains the frequency deviation in bounds. Without primary response, the grid frequency and therefore the whole grid would soon collapse. However, primary response is not able to restore the frequency back to its nominal value once it has been distorted. This is done by secondary and tertiary response, which is covered in the next sections.

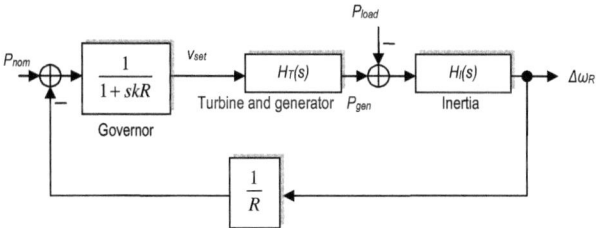

Figure 2.5: Generator with droop governor for primary response

2.1.2 Multiple generators and loads

So far, only a single generator was considered. The linear droop characteristic makes it easy to add multiple generators to a joint system. For steady-state examinations, the steady-state power-frequency characteristics of all loads and generators can simply be added (loads and generators separately). These accumulated characteristics can then be overlaid and the working point of the system can be seen from this (compare Figure 2.6). This is can be done since the losses in the power

Technical Background

lines connecting the resources account only for a small percentage of the whole power exchange and therefore are be omitted here.

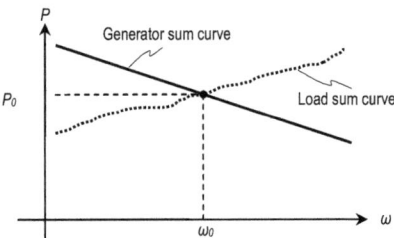

Figure 2.6: The steady-state grid frequency ω_0 can be graphically determined by overlaying the accumulated generation and load characteristics

On the basis of the same simplification, multiple generators are connected in the way shown in Figure 2.7 for the sake of dynamic modelling. The rotating masses of all generators are added to one lumped inertia model with

$$E_{rot,\,tot} = I_{tot}\,\omega_R^2 \quad \text{where} \quad I_{tot} = \sum_{k=1}^{n} I_k \,. \tag{3}$$

In a real power system, not all generators take part in primary response. In the British power system for example, the percentage of frequency responsive generation capacity is about 30 % [Pea07]. The remaining units just generate a fixed amount of power independently of the current grid frequency.

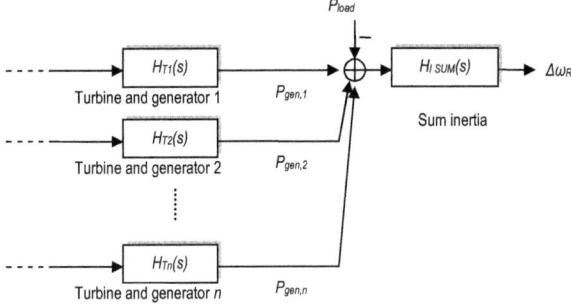

Figure 2.7: Construction of a lumped model for multiple generators

2.1.3 Generator for secondary response

The terms "secondary response" or "secondary control" refer to technical measures to bring the grid frequency back to its nominal value in case it has been distorted. Practically this is achieved by adjusting the power set-points of selected units when frequency errors occur for a longer while. In

Figure 2.8 the necessary modification of the generator control system is shown: in parallel to the proportional feedback for droop operation an integrator with a long time-constant (several minutes) is added. The long time constant is necessary so that the set-point adjustments happen only when errors are present for a while and the adjustment does not interfere with the fast operation of primary response.

In practice, generation units usually provide either primary or secondary control. The secondary control action frees the primary controller, so that it has again the full control range available for further primary responses. Only selected units take part in secondary response. They are selected on technical and economic considerations. The provision of control power is often financially very attractive for the power plant operator. Since the reaction times for secondary response are moderate, it is possible to make on-line biddings for control power provision.

The selection process is further complicated in interconnected power systems. An example for an interconnected power system is the "Union for the Co-ordination of Transmission of Electricity" (UCTE), an association of transmission system operators in continental Europe [6]. All member grids of UCTE are physically connected and have the same system frequency (which can be monitored online on the UCTE website [6]). In interconnected power systems, contracts are made in advance how much power is exchanged between the member grids during a day. The power flow between the member grids is caused by generation-demand imbalances and can be influenced by careful selection of the units for secondary response.

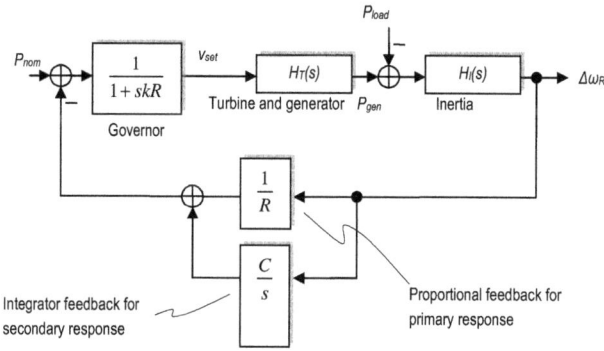

Figure 2.8: Generator model with governor for primary and secondary response

2.2 Electric loads

Since there are no restrictions for consumers when to switch on or off their electric loads, the modelling and forecasting of the demand side of the electric power grid is a wide-reaching discipline that has to take into account a number of different influences, namely social behaviour, climate, special public events and the like. One possible approach for modelling the demand side is the use of so-called "synthetic load profiles" [Fuenf00, Wer01]. Figure 2.9 shows the synthetic load profile for private households ("H0"). Synthetic load profiles are mainly used (and were initiated) by the power supplying industry and serve as reference load profiles for consumers who have no load measurement performed on site. Synthetic load profiles are the result of a statistical analysis based on representative samples from different consumer groups: households, shops (different groups for different opening hours) and industry (different groups for different working hours). They have to be scaled according to the year consumption of the customer. Each synthetic load profile consists of a set of day profiles with samples in 15 min intervals (96 samples/day) [Funf00]. The day profiles differ for each season of the year.

The load in the grid does not only change with time but also with frequency. The reason for this is that a part of the load is caused by rotating machines [Short07]. When the frequency drops, also the rotation speed decreases, and these machines use less power. Empirically it has been found [Kund94] that the frequency-dependence of the total load in the power system can be approximated by

$$P_{Load} = P_{Load,nom}\left(1 + D\frac{f - f_{nom}}{f_{nom}}\right). \qquad (4)$$

The factor $D = 1...2$ % degrades slowly over time since more and more electric machines are no longer directly connected to the grid but rather equipped with power electronics for torque or speed control, which separate power dissipation and grid frequency. For the European power grids (UCTE), $D = 1$ % is specified in [UCTE04a].

The frequency responsiveness of some loads in the system eases the burden of control actions lying on the generators selected for primary response. Nevertheless, the effect of this load-frequency dependability is lying in the area of 1 % [UCTE04a] and therefore rather small.

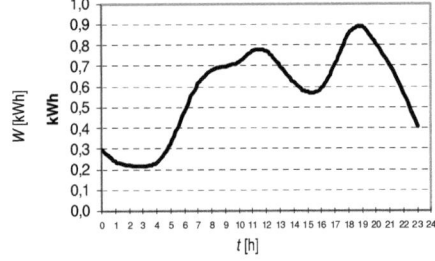

Figure 2.9: Load profile "H0" for household customers scaled to 5000 kWh/year [Funf00]

2.3 Differential equations for time-domain simulations

In the following it is described how the system block diagrams from the previous discussions can be turned into differential equations, which serve as a basis for time-domain simulations. The differential equations are further transformed into difference equations which can be directly used for time-domain simulation of generators in the power grid. These transformations are performed by applying the standard techniques of Laplace or Fourier transformation [Ohm06, p. 29].

2.3.1 Inertia

The Inertia of rotating masses and the power exchange between the grid and this inertia can be modelled in several ways. [Kund94] proposes a modelling approach with linearization in a working point (the problem itself is non-linear due to the fact that the rotation energy rises with the square of the rotation speed). For frequency-domain examinations it is advantageous to describe the whole system as linear. However, in this work time-domain simulations are performed, so there is no immediate need for linearization. Therefore, the more precise non-linear inertia model described in [Short07] is used.

For deriving this model, first the power balance is considered. Any surplus of generated power that exceeds the demanded power will – for reasons of energy conservation – be stored in rotation energy. The power surplus P_S therefore equals the derivation of the rotation energy:

$$P_s = P_{gen\,total} - P_{load\,total} = \frac{dE_{rot}}{dt}. \tag{5}$$

This is a first-order differential equation describing the relationship of power surplus and grid frequency. The following steps change this differential equation into a difference equation that can be used to calculate the grid frequency step by step on the basis of the known value in the last time-step. The derivation in (5) can also be written as differential quotient:

$$\lim_{dt \to 0} \frac{E_{rot}(t+dt) - E_{rot}(t)}{dt} = P_s. \tag{6}$$

For a numerical simulation, the time is divided in non-infinite time slices of length Δt. This factually means to leave away the limes in (6) as well as exchanging dt with Δt. By multiplying with Δt and adding $E_{rot}(t)$ on both sides, a formula for the *rotation energy in the next time-step* is found:

$$E_{rot}(t + \Delta t) = E_{rot}(t) + P_s \Delta t. \tag{7}$$

With (1), E_{rot} can be substituted:

$$\frac{1}{2} I \omega_R^2 (t + \Delta t) = \frac{1}{2} I \omega_R^2 (t) + P_s \Delta t. \tag{8}$$

Finally, for the grid frequency it is found, which is used in later simulations:

Technical Background

$$\omega_R(t+\Delta t) = \sqrt{\omega_R^2(t) + \frac{2P_s \Delta t}{I}}. \tag{9}$$

According to [Kund94], the inertial moment I can be approximated using

$$I = \frac{2HP_{G\ nom\ installed}}{\omega_{R\ nom}^2}, \tag{10}$$

where H is a grid-specific time constant varying from 2 to 8 s [Kund94]. $P_{G\ nom\ installed}$ is the total generation capacity (specified at nominal frequency) of all generators that are connected to this grid.

2.3.2 Primary and secondary response

For deriving the simulation equations for a generator with primary and secondary response governor, the block diagram in Figure 2.8 is considered as a basis. Following the modelling approach shown in [Short07], the transfer functions of governor, turbine and generator are all three put together to a simple first-order transfer function. This is reflected in Figure 2.10 and factually means that all higher-order terms (second and higher derivation) are neglected for reasons of simplification.

In the frequency domain, it can now be seen from Figure 2.10 that

$$P_{gen}(s) = \frac{1}{1+sT}\left[\delta(s)P_{nom} - \left(\frac{1}{R}+\frac{C}{s}\right)\Delta\omega_R(s)\right]. \tag{11}$$

For generators without secondary response, the constant C is zero.

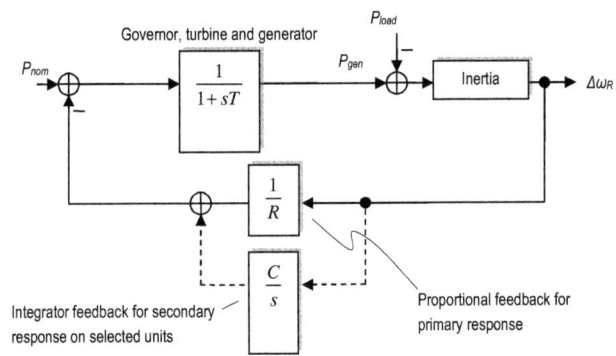

Figure 2.10: Generator model for deriving modelling equations for simulation

By multiplication with $1+sT$ on both sides it is achieved

$$(1+sT)P_{gen}(s) = \delta(s)P_{nom} - \left(\frac{1}{R}+\frac{C}{s}\right)\Delta\omega_R(s). \tag{12}$$

This equation is now transferred into the time-domain. Any multiplication of a system variable with s becomes a time derivation of this variable, dividing by s means integration over time.

$$P_{gen}(t) + \frac{d}{dt}P_{gen}T = P_{nom} - \frac{1}{R}\Delta\omega_R(t) - C\int_{\tau=-\infty}^{t}\Delta\omega_R(t)d\tau \tag{13}$$

Again, the derivation can be written using the differential quotient and, for time-discrete simulations, the limes can be omitted as well as dt can be substituted by Δt.

$$P_{gen}(t) + \frac{P_{gen}(t+\Delta t) - P_{gen}(t)}{\Delta t}T = P_{nom} - \frac{1}{R}\Delta\omega_R(t) - C\int_{\tau=-\infty}^{t}\Delta\omega_R(t)d\tau \tag{14}$$

Now, by applying simple addition and multiplication operations, a term for calculating the generation power *in the next time-step* can be found:

$$P_{gen}(t+\Delta t) = P_{gen}(t) + \left(\underbrace{P_{nom} - \frac{1}{R}\Delta\omega_R(t)}_{a(f-f_{nom})+b \text{ target power}} \underbrace{- C\int_{\tau=-\infty}^{t}\Delta\omega_R(t)d\tau}_{\text{secondary response}} - P_{gen}(t)\right)G\Delta t \tag{15}$$

Essentially, the current generation power is adjusted slowly step by step to meet the target power given by a term $a(f-f_{nom})+b$ (compare (2)) and, when C is nonzero, by a correction term that integrates the frequency error. $G = T^{-1}$ is the governor gain; it determines how fast the system reacts on changes in the target power. According to [Short07], a realistic value for G is 0.3 s^{-1}.

The target power obeys the linear droop characteristic $a(f-f_{nom})+b$. By comparison, the values for a and b can be found:

$$a = -\frac{1}{2\pi R}, b = P_{nom}, \text{ using } \Delta\omega_R(t) = 2\pi(f(t) - f_{nom}) \tag{16}$$

2.3.3 Example of sudden increase in load: simulation model and results

While the steady-state power-frequency characteristic of the grid is easy to construct by adding all individual characteristics of all resources in the grid, it gives only answer to the question "The power mismatch between generation and demand is x, what will be the frequency after the dynamic processes have come to a steady state?" In the previous section however the equations for a dynamic simulation of generators and loads in the grid have been derived. These enable answers to more complex questions such as "How does the system react on an increase in load of x MW?" Of course it has to be kept in mind that all equations presented are only modelling approaches for the real behaviour of the power grid system. Simulation results will only come close to the original behaviour, but always differ from it. The demand on the preciseness of the model is defined by the

Technical Background

application. Here, the model application is to verify and compare the principal functionality of different approaches for active loads that take part in primary or secondary response (see chapters 5 and 7). As long as the model of the power system is the same for all examined cases, the results are comparable. The model used is well-established and widely used (see [Kund94], [Short07], [Inf07]), therefore it is considered to be sufficient for the purpose of this work.

In order to illustrate the equations derived above, the example of a sudden increase in power demand shall be examined and the results derived from the model discussed. A sudden load increase can easily occur when a large consumer, such as an industrial electric oven, is switched on. The setup of the simulation experiment is depicted in Figure 2.11. The total generation capacity available in the small example grid is 30 GW, whereas only a small fraction of this is frequency-responsive (3 GW). In the simulation, this is reflected by two generator instances G1 and G2, one modelling all non-frequency-responsive generators (G1) and one modelling the generators with frequency response enabled (G2). On the demand side, there is a constant load L2 which is frequency-dependant according to (4). Additionally, the sudden increase in load is modelled by switching on the 1 GW load L1 at $t=0$. Inertias of all generators is reflected according to (3) using a lumped inertia model.

Before the additional load L1 is switched on, the grid is balanced. The grid frequency is at the nominal value 50 Hz, and the generation of 30 MW exactly matches the demand, which is also 30 MW at 50 Hz. As soon as L1 is switched on, a mismatch of 1 MW arises. The reaction of the example grid to this mismatch is shown in Figure 2.12 (primary response only) and Figure 2.13 (primary and secondary response).

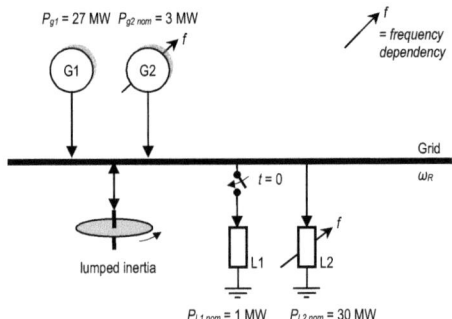

Figure 2.11: Setup for the generator drop simulation. G1 models constant generation in the grid. G2 models primary and secondary response generators in the grid. The load L1 is added to the system at $t=0$. The load L2 is constant over time but frequency-dependent.

First, the primary response shall be considered. As an immediate reaction on the increase in demand, power is supplied from the inertia (rotating masses). As a consequence, the rotating speed and with it the grid frequency decreases. The decrease in frequency has two effects: the frequency-dependant demand decreases (which can be interpreted as power supply from released demand), and, since this is not sufficient to account for the gap between generation and demand, the primary controller starts to increase the power generated. In Figure 2.12 this is shown as "control power from G2". The control power is the difference between nominal generation power and actual generation power. The increase of generation power relieves the inertia and this constrains the frequency decrease. Due to the governor delay, the system overshoots a little. The governor action comes to a halt when the increase of generation power plus the power from released demand account for the additional demand of 1 MW. Due to the droop characteristic of the governor, a final steady-state frequency deviation from the nominal 50 Hz of about 0.55 Hz can be observed. It should be noted that in a real system, not only one but multiple generators that do not operate at part capacity will increase their output. In this model, this reaction of multiple units is concentrated to a single generator response.

Second, the effect of the secondary control action shall be considered. It is depicted in Figure 2.13. Note the difference in time-scale to Figure 2.12 which is due to the slower reaction time of the secondary controller. In this case, the constant C in (15) was assumed to be 30 kW. The secondary controller initiates a slow but long-term increase of generation power to satisfy the increased demand. It not necessarily is active in the same generation unit as the primary controller, but in this simple model the same unit performs primary and secondary response. With the generation power finally increased according to the demand, the system comes back to a balance state where the power exchange with inertia goes back to zero, and also the released demand goes back to zero since the frequency is restored. In contrast to initial conditions, load and generation have increased by 1 MW.

The same simulation model will be used in Chapter 5. There, different types of frequency-responsive loads are added to the simulated grid and are compared within that framework.

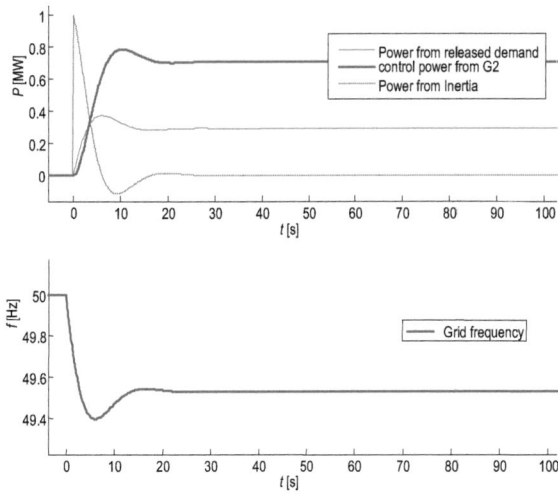

Figure 2.12: Simulation of a sudden load increase of 1 GW at $t=0$. Only primary response is simulated (no secondary response), therefore a steady-state frequency deviation can be observed.

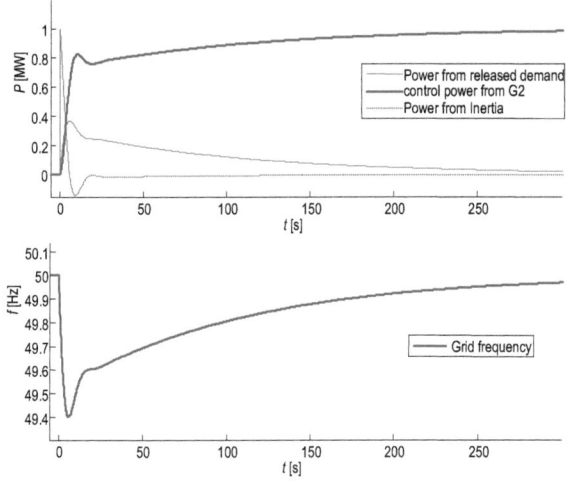

Figure 2.13: Primary and secondary response to the load increase at $t=0$. The frequency is restored to 50 Hz by the secondary control action. Note the difference in time-scale to previous figure.

3. Related work

This thesis bases on the results of earlier work in the area of demand side management (DSM). DSM, i.e. the modulation of power dissipation of electric loads motivated by energy savings and efficiency improvement, is a wide-reaching area. However, from the results of previous work described in the following, the conclusion is drawn to specialise in the specific direction of frequency-responsive loads. Related work in this specific area is therefore also reviewed, before the remainder of the thesis elaborates a new algorithm for frequency-response of demand-side energy storages.

3.1 Previous research in the context of this thesis

The main basis for this work was laid by the research project "Integral Resource Optimisation Network" (IRON) [Pal06, Stad05, Roe05, Kup06, Stad07, Kup08]. This collaborative Austrian project, with participating partners coming from university, energy service providers and grid operators, has studied the economic effects of "intelligent loads" and "virtual energy storages" in the power grid [Pal06].

"Intelligent loads" are electrical loads that can forecast their consumption behaviour for a certain time. The concept of "virtual energy storages" refers to the fact that many electrical loads incorporate (so far non-utilised) energy storage capabilities in the form of thermal or potential energy. The central idea of the project is that providing an infrastructure for distributed energy management and generally intensifying the information flow in the energy system will have positive effects on economic and energetic efficiency of the whole electric energy system [Pal06, p. 5]. It can be seen from electric storage heating schemes, which are classic examples of utilised energy storages in the demand side, that DSM can be useful for the operation of the power grid. However, storage heating is the only widely-implemented load management measure that can be found in today's power grids. Due to its very low energy efficiency, electric storage heating is generally retreating and will only remain, where electric energy is available very cheaply. So, the approach is to extend DSM to other types of loads and make the system more intelligent compared to the coarse grain technology of storage heating, where load management is done with local timers or with unidirectional activation and deactivation broadcasts [Kup08, p. 19].

Related work

Before looking more closely on the technologies needed for realising such an "intelligent" DSM system, the IRON project studied the macro-economic effects of such a system [Stad05, Stad07]. In the context of frequent electricity tariff hikes [Li03], which burden economies worldwide and result in higher inflation or economic cool down, load-management programs have to be of high interest for an economy, besides efficiency measures and the Kyoto Protocol targets. The reason is that such programs stabilize electricity prices and therefore reduce the negative effects of price spikes on the economy [Stad05]. With the model for electricity pricing shown in Figure 3.1 this fact can be illustrated.

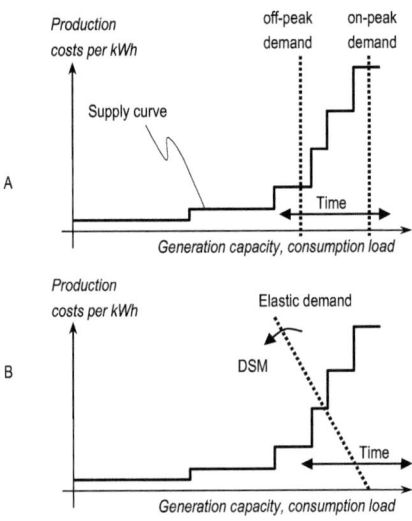

Figure 3.1: Inelastic (A) and elastic (B) demand. The intersection of supply and demand curve determines the energy price [Stad05].

The electricity price is depending on the amount of electricity demanded. The reason for this is that for a set of generation units, the production costs are not equal for all units but differ from unit to unit. While e.g. hydro power is cheap once the power plant is erected, power from coal or gas is more expensive. The cheaper generation capacities are scheduled before the more expensive ones. Thus, the electricity price does not linearly scale with the demanded amount but rather in an nearly exponential way (see supply curve in Figure 3.1). With energy consumers having a flat-rate tariff and not being aware of the capacity utilisation of the generator park, their demand curve in the diagram is a vertical line that moves on the horizontal axis over time (inelastic demand curves in Figure 3.1 A). With increasing demand, the energy price rises disproportionally, resulting in the above mentioned electricity tariff hikes on the wholesale energy markets.

Figure 3.1 B shows the result of the introduction of a technical infrastructure that makes unused load shift potential accessible and gives consumers the possibility to respond to price spikes easily in the short term without sacrificing comfort or services. Now, the demand curve has become elastic in the short term, meaning that the demand decreases in the short term when prices increase. This feedback mechanism stabilises the price situation on the energy market.

Inspired by this result, market models for integrating intelligent, short-term DSM solutions into the existing European and especially Austrian power market were studied. Four different approaches were examined in detail [Kup08, pp. 43 – 77]:

1. "Eco-Energy": Using the DSM infrastructure for an intensified utilisation of energy from renewable energy resources when it is available. Profits are gained from the low price of energy from renewables when the renewable resources are available.

2. "Loss optimisation": Using the DSM infrastructure to minimise the losses of electricity transmission and distribution. Profits are gained by the reduction of losses.

3. "Time-variable pricing": The dynamic of energy tariffs on the stock market is partly forwarded to the end consumer. Profit is gained by shifting consumption into times of low prices.

4. "Ancillary services": The DSM infrastructure is used to enable end consumers to provide ancillary services to the grid, especially imbalance energy provision (primary or secondary response).

It has been found that only the last two options, "Time-variable pricing" and "Ancillary services", are feasible in today's power market structure. The first two options fail to generate positive financial reward. However, provided that electricity prices keep increasing and costs for the DSM infrastructure fall, in future also the first two options might become realisable.

So, two main objectives for making use of "intelligent loads" and "virtual energy storages" remain. The first is the economically driven dispatch of storage capacities in order to gain profit from storing cheap energy for times of higher energy prices. This does only make sense in the presence of time-variable energy pricing. The storage scheduling strategy here is relatively simple because each single storage can be operated on an individual basis. Only the price information has to be broadcasted. This economic-driven dispatch approach can be divided into two sub-cases. In the simple case, the impact of storage activities on the energy market is small and prices are not influenced by DSM measures. This case has e.g. been studied by Nieuwenhout et al. [Nie06] for electrical storages (batteries). However, the outcome of this study is that the economic benefit is restricted due to the energetic losses in the battery systems (up-to-date battery technology assumed). The stock market price variations are not high enough to compensate these losses. This approach is only beneficial when battery technologies improve. Also optimistic estimations done in the IRON Project show, that only minor electricity cost reductions can be achieved by end-consumers reacting on stock market price variations [Kup06]. In the more complex case of the economic-driven dispatch approach, the impact of DSM activities becomes so high that it has significant influence on energy pricing (feedback phenomenon). Here, the costs of energy transfer over country borders play a major role. Not all countries in a common energy market will introduce DSM measures at the same time. If no artificial transportation costs or congestion charges are introduced, the benefits of an elastic demand curve in one country will immediately be exploited by the international market, so that no reduced energy price can emerge [Stad03b]. The DSM technology can go so far that automated agents take part in energy trading, offering, selling and buying energy portions at an on-line market place. Numerous studies have been undertaken in this area. One of the classical examples is the work of Ygge, Gustavsson and Akkermans [Ygge96]. They model the DSM problem by representing the participants using intelligent software agents which act in a computational market economy. Their simulation results reveal that this approach has a good scalability and achieves cost and energy savings.

The second objective for making use of "intelligent loads" and "virtual energy storages" is to enable the demand side to provide ancillary services such as imbalance energy provision to the power grid. By selling imbalance energy to the grid, profits can be made that are substantially higher than those gained from storing cheap energy and use it at peak-price times [Kup08, pp. 56 – 63]. On the technology side, this requires to schedule the DSM resources in such a way that a certain load profile is followed. In contrast to the first objective, for load profile following a collective DSM resource management is required that takes all available resources into account. Consequently, a considerably complex algorithm is needed that handles a large number of individual resources with different properties and is able to cope with statistical effects such as non-reachable resources or time-variant storage losses in virtual storages. An agent-based solution for this is proposed by Palensky [Pal01], who uses the agent paradigm and the concept of "intelligent loads" (see above) to achieve a certain overall load profile of a huge set of distributed loads that take part in a DSM program. Each DSM load is connected to a local software agent that communicates different possible alternatives of consumption behaviour of the local load to other agents. With the help of genetic algorithms, the most appropriate option among a large set of different solutions is selected. Finally, the agents controls the attached DSM load in such a way that the selected behavioural option is executed. This approach solves the problem, but it requires a real-time communication infrastructure where load agents come to a conclusion in a time horizon of second-range. Real-time energy management over a power-line communication infrastructure is proposed in [Trey04].

Whatever algorithm is used to schedule the DSM-resource behaviour, communication is needed between the individual nodes of the system. In agent-based approaches, Internet connectivity is usually assumed. However, due to the best-effort approach of Internet communication, no guaranties regarding availability, bandwidth and delay are given. It is indeed possible to subscribe to a Quality-of-Service scheme; however the costs for such a connection with guaranteed behaviour are usually too high for DSM applications. Under current market conditions, short-term DSM is only economically feasible if a low-cost communication medium is used and the DSM components are cheap by economy of scale [Kup08, p. 155]. As for any distributed control system, hard real time requirements are usually not avoidable for the scheduling of DSM resources, so the use of an inexpensive general purpose network with best effort transport bears a conceptual problem: real-time control using a non-real-time communication infrastructure. While no general solution for this dilemma exists, two techniques can be used to ease the situation [Kup07b]. One is to use synchronised clocks to trigger reactions at different places at a point in time that previously was arranged over the non-real time infrastructure. The second is to make use of implicit communication opportunities which are provided by the controlled process itself, in this case the power grid.

As discussed in the previous chapter, the power grid "broadcasts" the balance information to all resources in the system in the form of the grid frequency. The deviation of the grid frequency from its nominal value is proportional to the mismatch between supply and demand in the system. Therefore, it is attractive to make use of this ubiquitous information and base time-critical decisions within the DSM algorithm not on best-effort communication but on changes in the grid frequency, that can be seen as events on the implicit frequency communication channel. Since imbalance energy provision is currently the most promising market model for DSM applications as discussed above, and due to the availability of the frequency channel no need for a real-time communication infrastructure remains for this application, this work focuses on frequency-responsive loads.

3.2 Energy storage for neutralising load fluctuations

In a deregulated market environment for energy, not only the energy provision itself but also the ancillary services, which support the grid and are crucial for a stable grid operation, are assigned with a market value (and price). Frequency control is one of these ancillary services, which have only been provided by the supply side of the electrical power grid before deregulation and were a part of the integral task of energy generation. Now, that these services have an established market value on their own, new and unconventional solutions have the chance to emerge, that provide ancillary services independent of energy generation [Laz04].

One example for this is that the provision of control energy for frequency stabilisation can also be realised with energy storages. Lararewicz and Rojas propose a fly-wheel based storage system for this purpose [Laz04]. Fly-wheels are mechanical energy storages, where rotating masses are accelerated or decelerated by positive or negative electrical power using the dynamo-electric principle. As conceptually shown in Figure 3.2, the accumulated daily load is fluctuating fast on a second or even sub-second time horizon. These fluctuations are traditionally neutralised by varying the power generated from fossil fuels or hydro power. These generators have longer transients in the area of multiple seconds to multiple minutes [Laz04]. Energy storage is an appealing solution here since a surplus in power can be saved to neutralise a later lack in power.

Related work

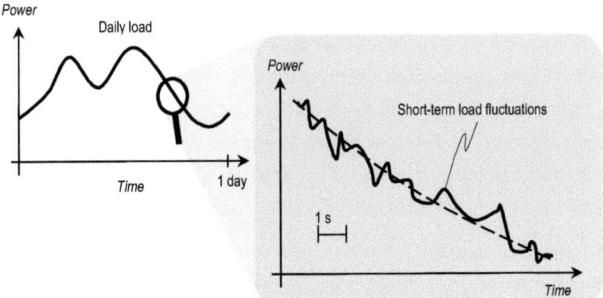

Figure 3.2: Load fluctuations in the daily load profile [Laz04]

By introducing fly-wheels into the grid, the existing energy storage capacity of the grid due to the rotating masses of the generators (and of some loads, see Section 2.2) is enhanced and therefore frequency stability is increased. According to [Laz04], by the application of an adequate amount of fly-wheel energy storage, generators would be unburdened from the duty of primary control, grid oscillations would be attenuated and even up to 15 minutes backup energy would be available, which would increase the grid reliability.

Similar considerations can be found in literature over the last decades, featuring different storage technologies such as superconducting magnetic storage [Ban90] or battery systems [Nie06]. However, the only storage system that is widely used so far is pumped hydro storage, where the economy of scale is applicable. In the context of market deregulation and increased need for ancillary services due to utilisation of renewable energy resources, this might change in future. The advantages of the storage approach are eminent, and storage technologies will become more mature.

The approach followed in this thesis is similar and has been proposed by Stadler et al. in [Stadt03a]. Instead of costly setup of new distributed storages in the grid, existing storage capabilities in the energy consumption processes (the DSM resources) can also be utilised. Here, the only costs are caused by the integration of the demand side storage into an embracing automation infrastructure which coordinates the activity of a large set of small resources, which thus act as a single storage. In this thesis, a new coordination algorithm for the utilisation of these demand side storages is developed.

3.3 Frequency-responsive loads

Load shedding, i.e. switching off groups of loads by the grid operator, is a state-of-the-art technique that is applied in emergency situations when the grid would otherwise get instable or break down. The automated activation of load shedding is often done on the basis of frequency information. The system frequency is used as one local indicator for the overall grid situation in protection systems that switch off loads in critical grid situations (see e.g. [Shok05]). However, this approach deals only with extra-ordinary grid situations where the energy provision to some consumers can be sacrificed (that is, completely switched off) in order to prevent larger blackouts. More intelligent approaches, where frequency-dependent adaptive load management is performed which is active during normal system operation as well as in emergency situations, are still subject to research and not jet widely implemented.

3.3.1 "GridFriendly^{TM} Appliance Controller": threshold-based algorithm

One approach that focuses on frequency-responsive loads is part of the large U.S. "GridWise^{TM}" initiative, where research in the area of smart energy grids is conducted [4]. In the corresponding subproject the "GridFriendly^{TM} Appliance Controller" (GFA Controller) has been implemented [Ham07, p. v]. This controller (shown in Figure 3.3) is supposed to be integrated in a large number of consumer products and performs frequency measurements. If the frequency drops below a certain threshold level, it reacts by reducing the connected load. Communication in this system is restricted to the system frequency channel only, resulting in restricted control possibilities.

Figure 3.3: The GridFriendly^{TM} Appliance Controller

The central idea of this approach is that a relatively simple frequency-response algorithm without the need for an additional communication infrastructure is implemented on a single chip (an FPGA-based solution is used in [Ham07]) that can be integrated in mass-market consumer products. By integrating the DSM intelligence into the end user product, a very cost-effective solution can be achieved. This is primarily due to two reasons: First, the interface between DSM hardware (here: GFA Controller) and end user equipment is an appliance-specific internal interface in the equipment itself. Double infrastructure for the DSM hardware, such as housing and power supply, can be avoided. Specifics of the controlled equipment can be fully utilised. Second, by providing a standard-chip solution, the economy of scale can be exploited.

Related work

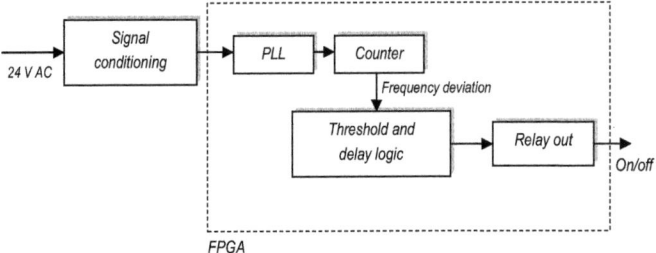

Figure 3.4: Block diagram of the functionality implemented on the GridFriendly™ Appliance Controller [Ham07, p. 1.9]

Details of the functions of this standard chip proposed by [Ham07] are shown in Figure 3.4. The alternating grid voltage is externally conditioned to a square wave signal, which is fed into the chip. This signal is again conditioned using a Phased Locked Loop (PLL) circuit. A counter with a fixed window time then determines the oscillation period. From this, the frequency deviation can be calculated. The following threshold and delay logic is the central peace of the DSM algorithm. It has to guarantee that not all "grid-friendly" loads switch off at the same time. Therefore, frequency thresholds and reaction delays are randomized within certain intervals. Each GFA controller has individual settings for these variables. Depending on the amplitude of the frequency drop, the controller reacts more or less quickly and activates the output signal. Depending on the load connected, the power is reduced. In the simplest case, the connected load is simply switched off. However, it is also possible that the load goes into an "economy" mode, in which it demands less power.

The project report [Ham07] deals with pilot installations of the GFA controller in 200 household appliances – 50 water heaters and 150 clothes dryers – and the corresponding installation and operation issues. The effect of these 200 loads on the power grid is clearly too weak to draw conclusions concerning the frequency-stabilisation effect of loads equipped with the GFA controller. As stated in [Ham07, p. vi], simulations for a higher number of loads are proposed for future work. A model of the GFA controller behaviour has been implemented as part of this work in order to compare the GFA approach to other algorithms. This is discussed in Section 6.2.5.

3.3.2 Approaches of continuous load modulation on grid-frequency basis

While the GFA Controller exhibits a threshold behaviour, which is motivated by ease of integration into the controlled appliance, many other proposals for frequency-responsive loads use continuous load modulation with the grid frequency. Continuous load modulation means that the power drawn by the loads in the system is modulated on the basis of a continuous relationship between frequency deviation and power deviation – which can be a linear function in the simplest case. The reason that many proposals use continuous modulation is that it results in less dynamic reaction on frequency changes and therefore reduces the risk of oscillations and instabilities.

Trudnowsky et al. assume a load modulation which is linear to the frequency deviation and study the effects of frequency control in the North American power grid [Trud06]. From their detailed simulations, they conclude that frequency-dependent load control has a number of benefits, which range from better system stability over more cost-effective transmission systems to an improved frequency control, which enables to reduce the number of generators providing primary and secondary control (Section 2.1.3). These generators can then operate under baseline load, resulting in more efficient plant operation.

While in [Trud06] it is only assumed that the load modulation is achieved using "intelligent loads", it is not explicitly mentioned how this DSM measure shall technically be achieved. Concrete proposals are given e.g. by Short et al. [Short07, Inf07] or Xu et al. [Xu07]. Both focus on thermostat-controlled refrigeration loads. The reason for this is that for continuous load modulation, a continuous interference with the consumer appliance is required, and this cannot be done with every type of electrical load. However, cooling and heating processes are predestined for DSM applications due to their capability of storing thermal energy, which makes their energy demand flexible in time.

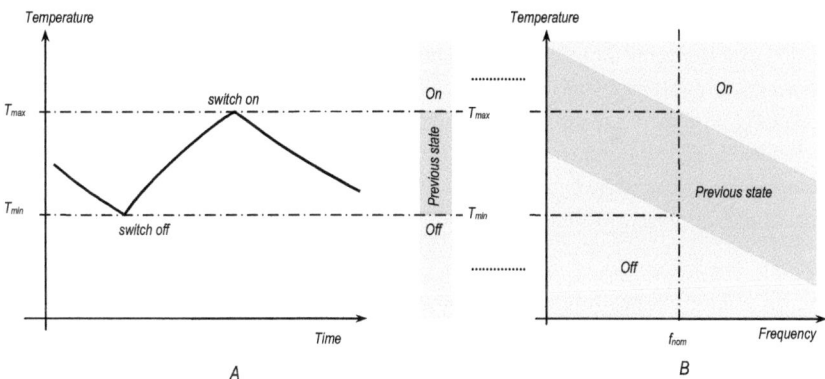

Figure 3.5: Thermostat operation (A) and the frequency-dependent setpoint change (B) proposed by Short et al. [Short07]

Figure 3.5 shows how load-frequency dependability is introduced by changing the temperature setpoint (or rather the upper and lower temperature thresholds) according to the grid frequency deviation. Most refrigeration loads (also air-conditioning systems, cold storage rooms etc.) are thermostat-controlled. The thermostat switches the compressor on if the temperature of the refrigerated space rises above an upper level und switches it off if the temperature falls below a lower level (Figure 3.5 A). This is the normal operation of the thermostat. The temperature approximately adheres to an exponential function (see Sections 4.3 and 4.5). For achieving a frequency response, both threshold temperatures are linearly shifted with the frequency deviation (Figure 3.5 B). The effect of this DSM measure is complex, since no closed mathematical formula can be derived for

39

Related work

the frequency dependency of the refrigeration load as a whole due to the non-linearity of the thermostat operations. Therefore, its effects on frequency control within a power grid can only be simulated; the British power grid is simulated in [Short07], Danish circumstances are assumed in the simulations presented in [Xu07]. Results support the conclusion of Trudnowsky et al. [Trud06]. As with the GFA controller, the approach of Short et al. has also been simulated in this work to compare it to the GFA controller and the new approach presented in this work.

3.4 Classification of demand side management

Apart from frequency-controlled systems, a number of research projects and test programs in the broader area of demand side management exist. The idea of demand side management (DSM) is not new. In the 90ies, it was already a widely discussed issue (see [Yau90] or [Smi94]). DSM approaches and measures have been and are used in numerous projects and schemes. Examples for these are demand reduction programs in the New York area [Law03], real-time pricing schemes in the entire U.S. [Bar04], or grid congestion relive in Japan by load management in small and mid-size enterprises on the basis of environmental data collected by sensors [Ish06]. In all these examples, the actual motivation and the technological approach are very different. Kiliccote et al. [Kil06] give a classification of different kinds of DSM, including motivations, technology design and operational aspects. The classification is intended mainly for DSM in large commercial buildings, but it can also be transferred to other application areas. Three different kinds of DSM are differentiated: efficiency measures (by change to better equipment), incentive-driven peak load management and event-driven demand response (Table 1).

Table 1: Classification of demand side management in the U.S. [Kil06]

	Efficiency and conservation (daily)	Peak load management (daily)	Demand response (dynamic event driven)
Motivation	• Conservation • Environmental protection	• Time Of Use (TOU) savings • Peak demand charges • Grid peak	• Price • Reliability • Emergency
Design	• Efficient shell, equipment & systems	• Low Power Design	• Dynamic control capability
Operations	• Integrated system operations	• Demand limiting • Demand shifting	• Demand shedding • Demand shifting • Demand limiting

The classification shows the two general approaches of on-line demand modulation (second and third column in Table 1): the first option is to set incentives for the energy end user to adapt the

40

consumption pattern. This is done by real-time energy tariffs (also called time of use tariffs) or peak demand charges. The second option is an event-driven approach, where the utility or grid operator tells the customer to react in case of a critical event, a procedure, which is agreed by contract before. Such an event occurs either in case of very high energy prices or in emergency situations.

In the classification of Kiliccote et al. the DSM system developed in this thesis falls into the dynamic response section. However, here the events notifications are not given by voice communication or Fax (as it is still the case in many DSM schemes in the U.S. [Kup06]) but rather by a sophisticated communication system relying on the grid frequency for real-time reaction and a non-realtime small-bandwidth coordination infrastructure.

In some places, DSM is actually present in the daily operation of the electricity grid since years. However, despite the high number of examples with successful DSM operation, in every project and in every application researchers and practitioners have to start factually again from zero. The reason for that lies in the strong individuality of network grid segments with their specific set of potential DSM resources. Differences are flattened when examining larger grid segments with more network nodes, but generally speaking the consideration of individualities is common in the majority of programs and projects. Additionally, with different motivations for using DSM in these programs (network stabilisation, peak load reduction, power quality improvement, energy trade etc.) also the technical measures and control strategies differ significantly. This finally goes along with separate legislative frameworks in different countries, resulting in a large number of special solutions. Thus, the question of what can be achieved with a given set of potential DSM loads currently cannot be answered. A range of possible benefits from the DSM operation in the given grid segment can be specified, but precise figures can only be gained from practical experiments, resulting in even more experimental DSM installations. In order to improve the comparability of results and the reusability of ideas, DSM concepts and control strategies, the basic properties common to all DSM resources have to be considered (Section 3.5).

3.5 Load shifting potential in Austria

A first assessment of the basic properties of DSM resources shall be derived from the work of Brauner et al. [Bra06]. In Figure 3.6, the total electrical energy demand of Austria is depicted over an exemplary winter day (load profile of Sunday, 11.12.2005). Brauner et al. have studied the DSM potential of Austrian households [Bra06, p. 56], both the result of this study and own estimations of the respective industry DSM potential are also depicted in Figure 3.6. The diagram allows comparing the fixed load with the flexible portions of the total load profile.

Related work

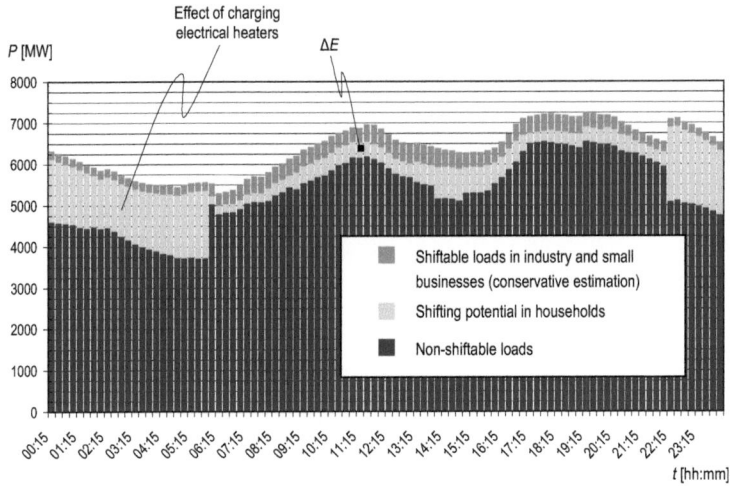

Figure 3.6: Load shift potentials in Austria compared to the total load of an exemplary winter Sunday [Bra06, p. 56]

The diagram suggests that the upper shaded parts of the consumed energy can be freely moved in time, e.g. to fill the valleys and flatten the demand profile. But this is not completely true. The gray areas are constituted by millions of single consumption processes, of which those in the upper two layers have (restricted) flexibilities concerning the timing of their consumption. So, a very small energy portion ΔE can be considered, which is marked in Figure 3.6. This energy portion is associated with an individual load process, which consumed power at 11:30 on the 11.12.2005 somewhere in Austria. This might have been a washing machine. It can be assumed without loss of generality that this energy could also have been consumed at 15:15 at the same day, if any motivation and means existed to reschedule the consumption process, but not at 18:00 for any reason.

Generally speaking, the flexible part of the load profile splits up in a large number of small energy portions ΔE, of which each single one can be moved in time, but not to arbitrary positions. Each ΔE is attached to its original position by a special kind of "rubber band" that allows it to move to certain new positions, mostly close to the original position. This flexibility and the detailed rules of which new positions are allowed and which are not allowed are individual properties of each consumption process.

From this example it can be seen:

(a) Adoptions of the load profile are possible, but only to a certain extent.

(b) The constraints result from a large number of individual process constraints.

(c) To achieve a required load profile adoption, a re-scheduling problem under the constraints given by each single DSM resource has to be solved.

Thus, independently of the actual motivation and aim of any DSM application, the task of finding the optimal resource allocation and dispatch has to be carried out. In order to do this efficiently in an effective way, it has to be examined how the individual constraints of power consumption processes interfere with each other, when adding up to the compound constraint for the adoption of the load profile. This examination is carried out in the following chapter.

4. Modelling resources for demand side management

Despite the wide-reaching associations that are related with the term "demand side management" (DSM), energy management on the demand side is merely a tool or an instrument. The utilisation of this instrument can have various different reasons. In the following, the instrument itself will be examined disregarding the motives for using it. This allows an objective investigation with a clear view on the basic behaviour of DSM resources (i.e. loads that can be influenced for DSM purposes). A model can be deducted that is applicable to virtually all potential DSM resources and will serve as basis for the control algorithm discussed in the next chapter.

4.1 Modelling approach using conceptual storages

The key concepts leading towards a generic description of DSM resource behaviour are load shedding, load shifting and energy storage. In case of load shedding, loads are simply switched off. Performing load shedding during on-peak consumption periods is a simple measure. Load shedding is unusually applied to unnecessary electricity consumption such as lighting in non-occupied rooms. However, in the light of sustainable energy usage, the contribution of load shedding to demand flexibility is questionable, since unnecessary loads should anyway be completely avoided. In some cases also useful loads are subject to load shedding, but here the user comfort is negatively influenced which rules this option out for frequent use during normal system operation. An example for this case is load curtailment in critical distribution grid situations, where consumer groups are disconnected from the grid in order to prevent the grid voltage to collapse. Consequently, only load shifting and energy storages at the consumer-side remain as feasible DSM measures. Energy storages in general can be realised by actual (i.e. "real") storages or by conceptual storages, i.e. load shifting. Here, the demand-side flexibility is constituted in the possibility to freely schedule a consumption process within certain restrictions defined by the considered application. Still, a reduction of the consumption is performed (e.g. during on-peak times), but the consumption is only delayed until more supply is available.

Load shifting can be performed with a large variety of different loads. Washing, cleaning, heating, chilling, pumping, all these electricity-consuming processes have, depending on the application, certain flexibilities in their time schedule [Bra06 p. 51 – 58, Stad03a, Kup07c]. In order to make effective use of the collective load shifting potential of a given set of potential DSM loads, each single process should be able to commit its maximum individual capability to the DSM framework. However, load shifting cannot be applied to every type of load. For example, the electricity demand of an elevator cannot be shifted in time. Nevertheless, for many loads a delay in operation of few minutes or more does not matter, e.g. electric hot water boiler. For some loads, even longer delays up to few hours can be acceptable, e.g. refrigerators and domestic dishwashers or washing machines. However, the load shifting capabilities for individual loads vary a lot and also depend on real-time factors such as e.g. changing hot water demand.

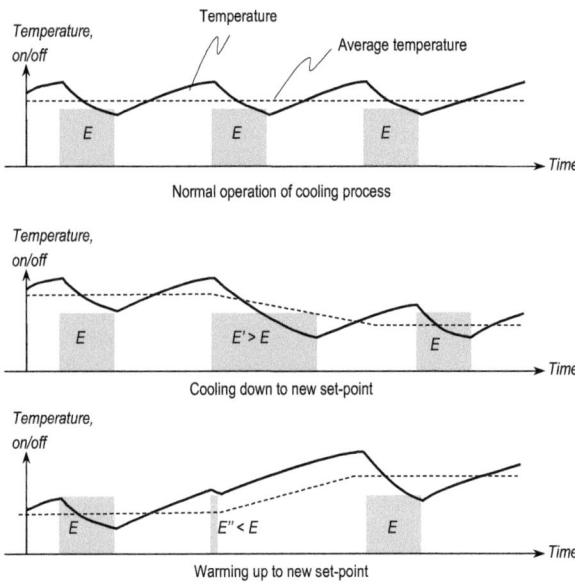

Figure 4.1: **Storing into and releasing energy from a two-point-regulated cooling process.**

An important group of electrical loads that can be used for load shifting are inert thermal process loads. Theses processes can be distinguished into heating applications (electrical heaters, domestic and industrial ovens, etc.) and cooling applications (air-conditioning, refrigerators, freezers etc). These account for more than 30 % of the electricity consumption of domestic households in the OECD [IEA03, p. 30]. Thermal processes have in common that they are able to store thermal energy due to the heat capacity of a room or a thermal-isolated box. The heat capacity can be utilized for shifting electrical energy consumption, if temperature set-points of the process allow a certain amount of variation. A typical thermal process is equipped with a two-point temperature regulator (thermostat), which keeps the temperature between an upper and a lower limit surrounding the actual set-point of the regulator (see Figure 4.1, top). Thermostat-control is widely used for its simplicity, low cost and reliable operation. Assuming energy E is needed to move the system temperature from the upper to the lower limit (without any distortions like opened refrigerator doors etc.), a decrease of the temperature set-point for a cooling process results in the system consuming more energy E' as in average (E) for the time it takes to reach the new temperature set-point. This circumstance is depicted in Figure 4.1, middle. When the original set-point is restored, the system consumes less energy (E'') as in average for a certain time period (see Figure 4.1, bottom). Thus, a load shift is performed by employing thermal energy storages. The advantage of thermal storage systems is that the possible storage time normally is very long or even unrestricted.

It should be noted that the problem is simplified in Figure 4.1 by assuming that once the new set-point is reached, it is again the energy E that is needed for keeping the temperature between the limits. E is actually a function of the set-point temperature (see Section 4.3.2).

Although the described measures are not factually storing electrical energy in a retrievable way, they can potentially be used to operate in the same way as energy storages and also to provide the same service as real electrical energy storages such as pumped storage schemes. Real electrical energy storages can of course also be used. In smaller scales, battery systems, especially Vanadium-Redox batteries [Sch05], or even flywheels [Heb02] are feasible. Nevertheless, costs for setting up such direct storages can be significantly higher than making use of existing consumption processes. Moreover, large-scale pumped hydro storage cannot be realised everywhere because of geographical restrictions.

The power consumption pattern of an electrical load that takes part in a load shift action can be described as the superposition of the unmodified process and a storage pattern. The fact that load shifting can conceptually be described as storing and releasing energy is exploited by the generic description model presented in the next section.

4.1.1 First approach for a generic model of load shifting

When implementing a self-controlled system performing the previously discussed measures of load shedding and load shifting, all (distributed) participating loads can be seen as resources. Operating the system does basically mean to solve the problem of optimal resource allocation and schedule. A preferably simple and consistent description of the resources involved is crucial for the dispatch algorithm to be efficient and flexible.

The first step towards such a generic model is to describe all DSM techniques of load shedding, load shifting and utilisation of real electricity storages as special cases of a general class of conceptual electricity storages. Such a conceptual energy storage can be charged and discharged and has an internal energy state s that changes with charge and discharge. The letter s will be used for the energy level in DSM resources throughout this thesis. The value range of s is restricted, which reflects the fact that the amount of energy that can be stored in the storage is not infinite. When the storage is charged, it consumes power, and s increases. When it is discharged, it acts as a power source, and s decreases. The variable s can have positive and also negative values, depending on how $s = 0$ is defined. The state $s = 0$ should reflect the normal energy state of the resource.

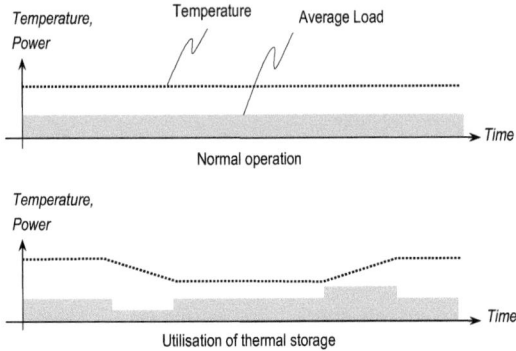

Figure 4.2: Utilisation of a thermal storage. The temperature is allowed to vary in certain bounds.

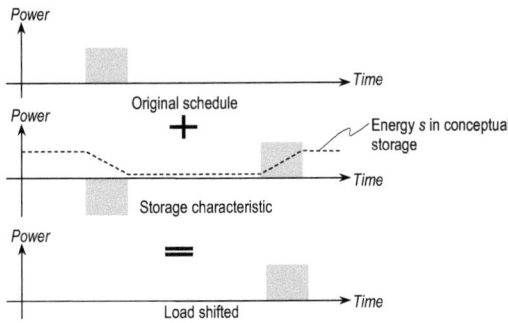

Figure 4.3: Classic load shifting can conceptually be described as the superposition of the original schedule and a storage characteristic.

This simple storage model is valid for three classes of resources: firstly, *real electricity storages* are energy storages without question and therefore fall under this general class. Secondly, concerning *other types of energy storages*, the example of inert thermal processes has been discussed already. Although loads incorporating such process are not able to return stored energy directly as electricity, they increase and reduce their electrical energy consumption when charging and discharging the thermal (or any other kind of energy) storage. As outlined in Figure 4.2, this behaviour can be described as the superposition of a normal load profile (constant or shaped) and a storage characteristic. Finally, in case of classic load shifting (e.g. delayed dishwasher operation) the process can be

Modelling resources for demand side management

described using a *hypothetical storage*. This hypothetic storage supplies energy at the time of the original load schedule and re-charges at the time of the delayed schedule (see Figure 4.3). In both cases the storage is described by the maximum energy it can store (s_{max}) and the maximum storage time (t_{store}). In this context, load shedding can then be described as a hypothetical storage with $t_{store} \rightarrow \infty$.

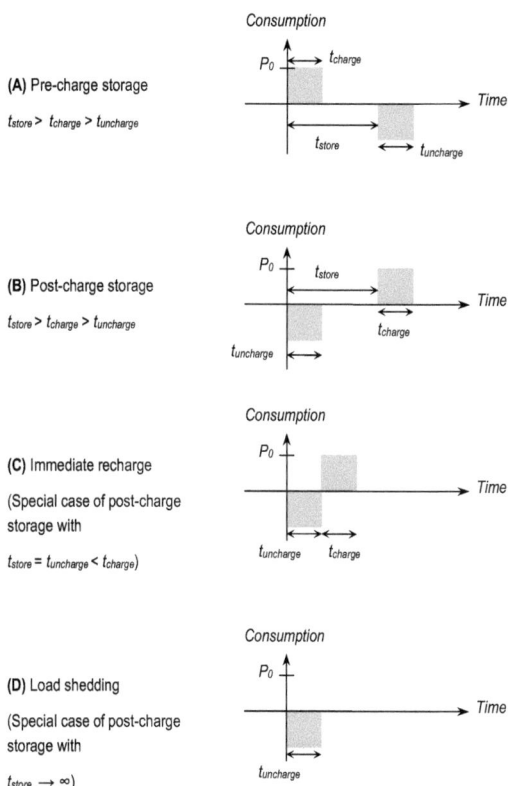

Figure 4.4: Two basic storage characteristics A, B and special cases C, D.

So, it is possible to describe all options for DSM measures using superposition of original load profiles and storage patterns, independently of whether real storages are involved or not. In the context of this work it is useful to define (a preferably small number of) basic storage characteristics, which can be used by DSM application algorithm. Figure 4.4 gives an overview of these basic storage characteristics. Generally, pre-charge and post-charge storages have to be considered. A pre-charge

49

storage charges before it can release energy. A post-charge storage charges *after* it has released energy. The latter option might seem to disagree with common sense, but it has to be kept in mind that these storage patterns are only meant to be superposed to a power consumption profile of an electrical load. As far as a real storage exists in the process, it will be operated at a certain set-point from which variations in two directions are possible. Thus, it actually makes sense to discharge before charging. The load shift in Figure 4.3 is also an example for a post-charge storage.

Besides the two basic types of pre- and post-charge storages, some special cases can be derived from them. For many applications, the charging period starts immediately when the discharge period has ended (post-charge). This is always the case when a charging process, such as warming up of a water boiler, is interrupted by a DSM measure. As soon as the interruption has ended, the process resumes in charging its internal energy reservoir up to its predefined maximum level. This is the "immediate recharge" case as shown in Figure 4.4 C. Additionally, as mentioned above, load shedding can then be described as a post-charge storage with $t_{store} \to \infty$ (Figure 4.4 D).

Having mapped the different options of load management to conceptual storage characteristics, each DSM resource can be modelled with an individual set of static parameters

$$\{P_0, t_{charge}, t_{uncharge}, t_{store}, t_{nostore}\}, \tag{17}$$

where

P_0 the power amplitude,

t_{charge} the time to charge the conceptual storage

$t_{uncharge}$ the time to discharge the conceptual storage (in some cases this time is smaller then T_{charge} due to losses in the process)

t_{store} the storage time

$t_{nostore}$ the minimum time between two storage operations (not depicted in Figure 4.4)

More complex procedures, e.g. those where different power amplitudes are involved for the charging and discharging, can be described by superposing multiple scaled instances of the basic prototypes. With this technique it is even possibly to adequately approximate non-rectangular profiles.

4.1.2 Concept of energy packets resulting in two dimensions of freedom

Peak load reduction as DSM application will be taken as an example for the following discussion, but the general concept can also be applied to other DSM applications, such imbalance energy provision and other auxiliary services.

Storages should be charged in times of high electricity availability and release it in times of peak demand. This is valid for real storages such as pumped storage schemes as well as conceptual storages as described in the previous section. Conceptual storages are available in large numbers, but they are distributed and have restricted individual storage capacities. Furthermore, many distributed storages have storage times less than 1 h (especially those of the immediate-recharge type), which appears to be one reason that utilization of distributed storages was regarded as in-effective in the past. However, if it is possible to gain DSM access to a significant number of distributed storages, the collective (because networked) storage capability is considerable.

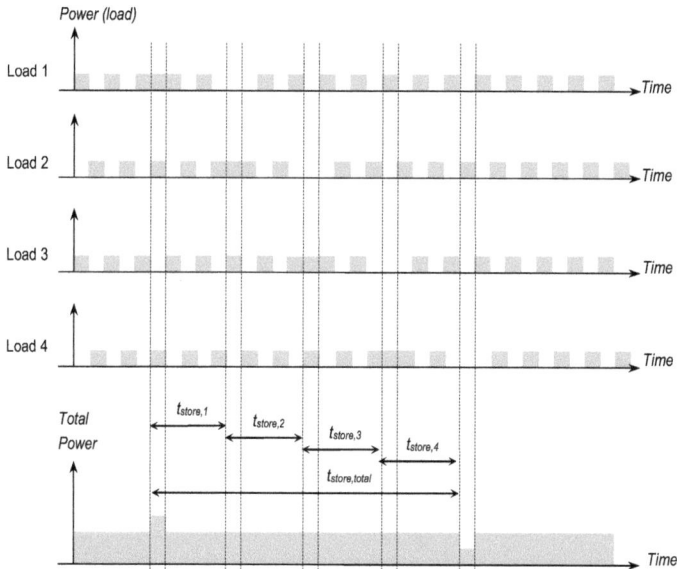

Figure 4.5: Increasing the storage time by utilizing multiple conceptual storages: four individual load shifts (top) accumulate to one long shift operation (bottom). Using this technique, the problem of restricted storage times can be solved.

Each conceptual storage i can hold the energy

$$s_i = P_{0,i} \cdot t_{charge,i} \qquad (18)$$

s_i can be seen as an *energy packet* of a certain size, which can reside in a conceptual storage for a certain time $t_{store,i}$. If a single (conceptual) storage is not able to hold an energy packet long enough, it can be transferred to other storages after the storage time – of the initial storage – has expired. Figure 4.5 shows a simplified example of four different duty-cycled consumption processes (e.g. refrigerators or air-condition systems). The first process stores additional thermal energy to release it (by reducing its power consumption) after $t_{store,1}$. At the same time, the second process loads its thermal storage to release the energy after $t_{store,2}$. This is repeated for processes 3 and 4. The total effect of this schedule is the same as that of a single storage with

$$t_{store,total} = t_{store,1} + t_{store,2} + t_{store,3} + t_{store,4}. \tag{19}$$

The serial connection of (conceptual) energy storages is outlined in detail in Figure 4.6. An energy packet is taken by load 1. The internal energy state is increased by this. After the storage time $t_{store,1}$ the resource has to release the packet again and return to its original energy state. In order not to "lose" the stored energy, it is conceptually handed over to another load, which takes the packet in the very same moment it is released by load 1. Finally, load 2 has to release the packet after $t_{store,2}$ as well. Here, another load could jump in, or the storage process is finished.

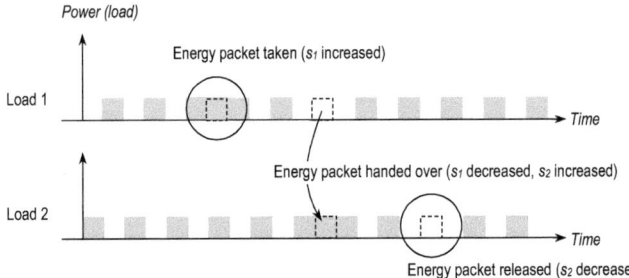

Figure 4.6: Coordinated load shift operation described using the concept of energy packets. Energy packets can reside in DSM resources (reflected by a higher energy state s_i) and can be transferred between resources without losses.

Energy losses occur in consumption processes due to non-ideal energy conversion (e.g. from electrical to thermal) within the process, but even before due to the transmission over non-ideally conducting transmission lines. (Please note that effects of losses are not shown in Figure 4.5 and Figure 4.6.) However, the transfer of energy packets from one process to the next does not cause additional losses. This is due to the fact that energy is not actually flowing from one process to the other process nor is electrical energy factually stored in a releasable way. In fact, just a subtle coordination of DSM resources is applied in order to achieve a modulation of the overall power consumption that is equal to that achieved by a real storage. It is worth noting that in this aspect conceptual storages have an actual advantage over real storages. Due to the ideal transferral, packets can be stored in virtually endless chains of DSM resources (the same resource might be involved in such a chain for several times). In contrast to that, if energy portions would be handed over from one battery to the next, the previous always charging the next and so on, the remaining energy portion would become smaller and smaller very quickly.

With the new concept for combining DSM resources (a) larger capacities by means of parallel operation and/or (b) longer storage times by means of serial operation can be achieved. With a large number of resources, any legal combination of both options is possible. The following consideration defines the term "legal combination":

If all resources have the same properties s and t_{store}, then all achievable combinations would be given by (20).

$$s \cdot t_{store} = \text{const.} \tag{20}$$

But since each individual resource i has its own set of $\{s_i, t_{store,i}, t_{nostore,i}\}$, only the summation of all $s_i \cdot t_{store,i}$ is constant. The term $s_i \cdot t_{store,i}$ describes the storage potential of an individual conceptual storage with regard to its restricted storage time. Using this term, the DSM potential of resources can be compared. The objective of shifting the energy E from t to $t+\tau$ can be achieved in two ways using two basic different resources with different s_i and $t_{store,i}$. First, each storage is able to store ½E for τ or second, each resource is each able to store E for ½ τ. Both resources have the same storage potential ½$E \cdot \tau$.

4.2 Case study: following required load profile

After the individual DSM resource has been modelled, it is now possible to examine the way many resources can be operated together in order to achieve significant impact on the overall power consumption profile. The intention of the following case study is to give an example application for the model derived so far and to gain insights about the restrictions of the model in the context of this application. For this case study, an optimal schedule for distributed storages shall be generated on the basis of the DSM resource model. The term *optimal schedule* is used in a specific way here. An objective is assumed that shall be achieved by the DSM resource collective. The collective will not be able to meet the objective perfectly under any circumstances. For an optimal schedule the distance to the objective (according to a certain distance metric) is mathematically minimized. By applying mathematical optimisation techniques to the problem at this point, one strong advantage is gained: the behaviour of a system consisting of elements that are modelled by the derived DSM resource model can be studied without having to deal with any kind control algorithms specially developed for the given application, that would add their specific properties and restrictions to the case. In this sense, the mathematically optimal solution is "neural" and the fully determined by the basic model used – which shall critically be examined here.

4.2.1 Optimal scheduling of distributed storages

A large variety of choices for the utility function exist. The utility function can for instance be individual energy cost reduction in the presence of time-variable energy prices. In this case, each individual storage does its best to reduce consumption in high-price times. A storage with arbitrary storage time examines the anticipated price curve and searches for the lowest and highest point. Then it charges at the minimum and discharge at the maximum of the price curve. Alternatively, a resource with restricted storage time searches for the highest slope in the prospected price curve, since here the maximum gain can be achieved in the shortest time. However, this objective can be achieved without coordination between individual resources and therefore it is not a preferable example in this context.

More sophisticated resource scheduling is needed for another kind of objective: load profile following, which shall be realised here. The idea is to control all distributed resources as they were one large single storage. This concept is somewhat similar to the concept of virtual power plants, where (primarily) distributed generators are controlled as a single unit. The chance for the development of such virtual power plants has first been identified by Awerbuch and Preston [Awer97]. However, in the approach described in this work, the focus is on energy storage on the consumer side and not on generation. A large collective of DSM resources can have considerable capabilities and can e.g. be used for peak-shaving. The collective behaviour is controlled by defining a load curve that is followed (as good as possible) by the distributed resources. This is particularly interesting in the context of demand charges, where a large percentage of the energy bill is constituted by costs for the largest monthly or yearly peak demand (see [Stad06]). Costly peaks can be avoided by superposing them with a negative peak in the load profile of the DSM resource collective.

The objective-driven scheduling of DSM resources modelled as storages can be described as a linear optimisation problem. This optimisation approach can particularly help understanding the way resources are used to achieve certain overall consumption behaviours. Especially for a set of resources with strongly varying parameters, it is not at all obvious how these storages should be scheduled to form e.g. a sinus-shaped load profile. So, the task of finding a solution by hand in order to derive a computing algorithm from this is actually not a simple one. Using the linear optimisation technique, an optimal scheduling solution can be found for the given problem, and from this solution two important aspects can be seen: first, how close the result for a restricted set of resources can come to the requirement, and second, how the resources can be scheduled so that the result can be achieved.

4.2.2 Formulation of the linear optimisation problem

As outlined in Section 4.1.1, DSM resources can be modelled by the linear superposition of basic storage characteristics. Consequently, it is possible to translate the scheduling problem of DSM resources into a linear optimisation problem. In order to avoid the following discussion becoming unnecessarily complex, the optimisation problem shall only be discussed for resources with arbitrary storage time t_{store}.

An optimisation problem consists of *variables* to be optimised, an *utility function* describing the relationship between variables and the objective metric and *constraints* that define additional relationships between the variables and restrictions for the value ranges of individual variables. In this case, a linear mixed-integer problem can be formulated. This means that the variables in the problem can be both integer and real. When solving the problem, it is first searched for an optimal real solution that suits all conditions. After that, the so-called branching process is executed where for each variable the optimal next integer solution is searched recursively [Koch04].

The general approach used here is to search for optimal values of a set of variables where each variable determines the presence and amplitude of a storage characteristic for a certain resource and at a certain point in time (the problem is seen as time-discrete). The optimum is defined by a metric that compares the required sum function with the achieved one. A number of constraints have to be formulated to restrict the solution in regard to the electrical and physical constraints of the problem.

First, the variables used in this problem shall be discussed. Since a time-discrete problem is analysed, variables can be defined as series where each term stands for one time slot. Although series are infinite (as is the time in the real problem), calculations or optimisations respectively can be restricted to a finite section of the series. The variables are coupled by constraints that also will be subsequently outlined.

DSM resources can either be charged or discharged, resulting in a positive or negative contribution to the load profile. While a charge or discharge process may last over several time slots, a single issue time slot can be identified at which the process is scheduled. Two series $c_{n,i}$ and $d_{n,i}$ can be defined, so that

$c_{n,i} = 1$ for n is the issue time slot of a charging process for resource i,

or 0 otherwise, (21)

and

$d_{n,i} = 1$ for n is the issue time slot of a discharging process for resource i,

or 0 otherwise. (22)

It is useful to split charging and discharging into two series, so that the total number of charging processes per resource m can easily be restricted by expressions like

$$\sum_{\forall n} c_{n,i} < m \qquad (23)$$

without having to operate with absolute values. Nevertheless, a single series

$$a_{n,i} = c_{n,I} - d_{n,i} \qquad (24)$$

can also be defined, that includes both charging and discharging information. In $a_{n,i}$, $c_{n,i}$ and $d_{n,i}$, only one element is non-zero for each charging or discharging process. This can be seen as an impulse triggering the whole process as shown in Figure 4.7.

Modelling resources for demand side management

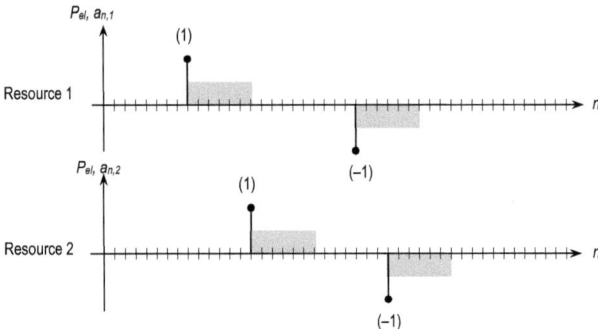

Figure 4.7: Relationship between the series $a_{n,i} = c_{n,i} - d_{n,i}$ and the actual charge and discharge processes.

Each resource has its individual charge profile $f_{n,i}$ and discharge profile $g_{n,i}$. These profiles can also be described as series. The convolution of the impulse series $c_{n,i}$ and $d_{n,i}$ with these time-discrete load profiles $f_{n,i}$ and $g_{n,i}$ result in the actual (time-discrete) power consumption $p_{n,i}$ of the resource n as described in (25). This equation is actually a linear constraint equation to the optimisation problem since it constraints $p_{n,i}$ in regard to the previously defined variables.

$$p_{n,i} = c_{n,i} * f_{n,i} - d_{n,i} * g_{n,i} = \sum_{k=-\infty}^{+\infty} c_{k,i} f_{n-k,i} - \sum_{l=-\infty}^{+\infty} d_{l,i} g_{n-l,i} \qquad (25)$$

When calculating the convolution, the sum index has only to cover those terms that are non-zero. In the following discussion it will always assumed that all processes are fully scheduled in the time interval $[0, T_{end}]$ with no process overlapping the borders of this interval. In this case, the convolution can be calculated as shown in (26) and thus has a finite number of terms.

$$a_n * b_n = \sum_{k=0}^{T_{end}} a_k b_{n-k} \qquad (26)$$

Furthermore, the energy level $s_{n,i}$ of resource n can be defined as

$$s_{n,i} = \sum_{t=0}^{n} p_{t,i}. \qquad (27)$$

To sum up the previous variable definitions, the impulse series $c_{n,i}$ and $d_{n,i}$ are the actual unknown variables defining the schedule of charging and discharging processes that shall be determined by the optimisation. Power $p_{n,i}$ and energy level $s_{n,i}$ are derived from them and are needed in the utility function and further constraints respectively.

The utility function subtracts the achieved overall load profile $p_{tot\,n}$ from the required profile $p_{req\,n}$:

$$\sum_{n=0}^{T_{end}} \left| p_{tot\,n} - p_{req\,n} \right| = \sum_{n=0}^{T_{end}} \left| \left(\sum_{i=0}^{N} p_{n,i} \right) - p_{req\,n} \right| \to \min, \qquad (28)$$

(where N is the total number of resources). The absolute value of the difference is used in order to keep the problem piecewise linear (it is possible to translate (28) into a set of linear equations with restricted index ranges [Koch04]). From an application perspective, the square of the difference would be preferable, but in that case the problem cannot be solved by solver for linear problems.

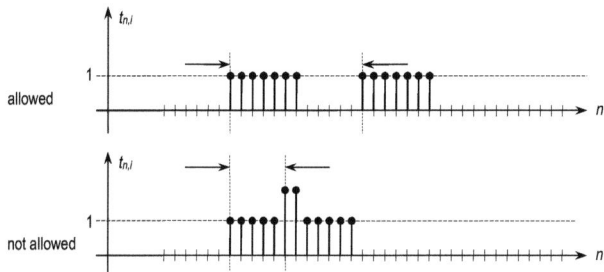

Figure 4.8: Avoiding that two charge or discharge processes are scheduled too closely by restricting the amplitude of the test series $t_{n,i}$ (see (31))

Further constraints of the problem are

- restricted storage capacity:

$$S_{min,i} \leq s_{n,i} \leq S_{max,i} \qquad (29)$$

- restricted charge and discharge frequency: the impulses in $c_{n,i}$ and $d_{n,i}$ must not come too close to each other (refer to $T_{nostore}$ in (17) on page 50). The resource may need some time to settle on a certain energy level until the next set-point change can occur. A possibility to express this in a linear equation is to make use of a distance series

$$\delta_{n,i} = 1 \quad \text{for} \quad n = 0 \dots T_{charge} + T_{nostore}, \quad 0 \text{ otherwise,} \qquad (30)$$

and to restrict the amplitude of $t_{n,i}$ (the convolution of distance series and impulse series) as outlined in (31) and Figure 4.8. Depending on the restrictions of actual resources it may be necessary to define separate δ series for charging and discharging or even to formulate more complex constraints.

$$t_{n,i} = (c_{n,i} + d_{n,i}) * \delta_{n,i} \leq 1 \qquad (31)$$

4.2.3 Solving the optimisation problem

Mixed-integer programs can efficiently be solved using a number of different algorithms [Acht04, pp. 1–2]. In the course of this work the open-source ZIMPL/SCIP framework [Koch04, Acht04] developed at Zuse Institute of TU Berlin was used. The advantage of this approach is the integrated description language for linear mixed-integer problems ZIMPL (Zuse Institute Mathematical Programming Language) that allows for specifying the problem in a human-readable form using sets, parameters, variables and constraint equations.

ZIMPL translates the program text into a matrix representation of the problem with a finite number of additional constraints. Depending on the size of the problem, the matrix can become rather large. For details on the matrix format, refer to [Acht04]. This problem representation is then passed to the solver SCIP (Solving Constraint Integer Programs). While simple problems can be solved within a few milliseconds, more complex optimisations can take several hours on a desktop computer to solve. For the given resource scheduling problem, the solving time is rising with the number of time steps and the number of resources. Especially the diversity of the resource parameters has a disproportionate high influence on the solving time.

The course of mathematical programming with ZIMPL is, as every programming process, not a straight procedure but a rather iterative programming and debugging process. Errors in the constraint formulas can easily result in an unsolvable problem, a fact that only becomes obvious after the solver tried to find a solution for some hours. It was found to be good programming practice avoiding complex constraints and rather defining more variables and coupling them with simpler constraints.

The optimisation result is a text file listing the optimal values for each variable. This file can be parsed in order to transform the data into yet another representation for additional processing stages. In the course of this work a set of problem-specific tools was developed for visualisation purposes. E.g. a scheduling diagram such as the one depicted in Figure 4.9 can automatically be generated from the solution file.

4.2.4 Solution to an example problem

In Figure 4.9 an example for optimal load profile following is depicted. Twenty thermal processes, A01 to A20, with similar properties (arbitrary storage time, equal storage capacity as well as equal charge and discharge times) were assumed in this example. The number of resources is chosen very low for better clarity here, realistic numbers are > 100. Nevertheless, similar results can be achieved with other storage types, e.g. with restricted storage time, using a slightly different mathematical description. The comparison of specified and achieved load curve is shown at the top. Below that, the charge/discharge schedule for a subset of the twenty resources is depicted. Not shown in Figure 4.6 is the actual power consumption of the loads used as DSM resources, which would simply add up to a constant offset.

The main outcome of this case study is that it is possible to combine the individual load profiles of resources in such a way that any required sum profile is followed as good as possible. This underlines the practicability of the modelling approach. However, since the so far presented case study is only realised in the framework of the model and does not take into account real-world data, it does not in any way validate the model. In fact, the discussion in Section 4.3 will show that a model refinement is necessary.

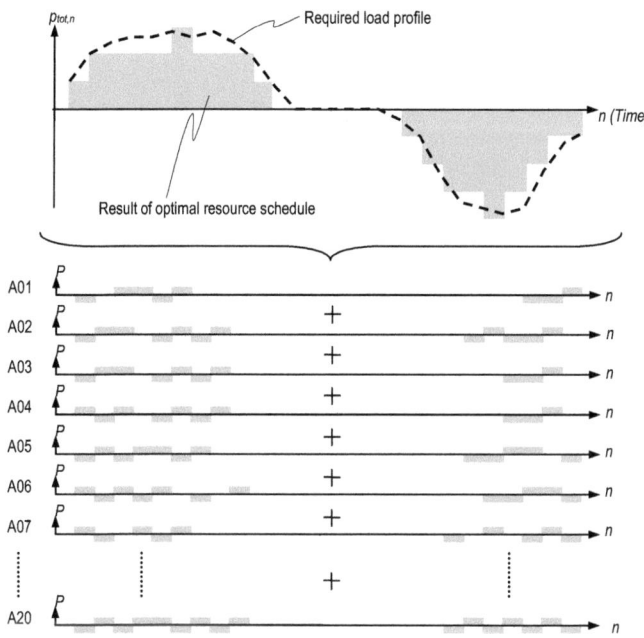

**Figure 4.9: Resources follow a required load profile (shown above).
The resource schedule in form of individual resource activities is shown below.
The achieved total load profile is the sum of all individual profiles.**

A simple model for the DSM resource behaviour has been found that describes resources as energy storages with an internal storage state s_i. The power consumption p_i of the DSM resource i can be calculated as the superposition of the power consumption during normal operation c_i and the power components caused by charging and discharging the (real or conceptual) storage:

$$p_i = \frac{ds_i}{dt} + c_i \qquad (32)$$

For the energy state function s_i, a number of conditions apply. The value range is bound, and the frequency of changes can also be restricted.

4.3 Model refinement for inert thermal processes

The model deducted so far is not very detailed and therefore simple. It is only based on the principle of real or conceptual energy storage in resources and does not account for any other effects than that. In particular it does not account for losses that depend on the internal energy level s. It is difficult to derive reasonably accurate generic models for the consumption behaviour of potential DSM loads due to the individuality of each single process and their stochastic behaviour. However, one particular kind of consumption process, which can be seen as a class of appliances playing a major role among DSM loads, can be modelled using a simple electrical model, which describes its behaviour more detailed than the model discussed so far.

4.3.1 Equivalent electrical model for a thermal process

Electrical heating and cooling applications in the domestic and industrial sector have one important property in common: electrical energy is primarily needed to replace the thermal energy lost due to intentional or un-intentional leakage in thermal insulations. This is true for both heating and cooling processes. A simple but general model for such different applications as supermarket refrigerators, chilled water or beverage vending machines, air-conditioned offices, heat pumps, cold storage rooms, boilers and many others can be derived from this observation. These systems consist of three elements:

1) An energy converter that transforms electrical energy in thermal energy (heater, chiller, heat pump etc.)

2) A place/area/room/compound whose temperature is different than that of the surrounding due to the influence of 1)

3) The bounds and isolation of this compound that allows heat exchange with the surrounding to a certain extend

In the model discussed here it will be assumed that the temperature is constant over the volume of the compounded area. The behaviour of the system can elegantly be described using an equivalent electrical circuit. By comparing the definitions of thermal and electrical variables, the following equivalent electrical elements can be found (see Table 2):

Thermal capacitance:

$$C_{th} = \frac{dQ}{dT} \quad (Q\text{: heat quantity, } T\text{: temperature}) \tag{33}$$

Thermal resistance:

$$R_{th} = \frac{T}{P_{th}} = \frac{T}{\frac{dQ}{dt}} \qquad (P_{th}\text{: thermal power}) \qquad (34)$$

This results in the simple electrical circuit depicted in Figure 4.10, which models the behaviour of the thermal system. In this circuit, C models the thermal capacity of the thermally insulated room; R_L models the thermal leakage due to non-ideal insulation.

Table 2: Thermal and equivalent electrical variables

Thermal	Electrical
Power $P_{th}=Q/t$	Current I
Temperature T	Voltage U
Resistance R_{th}	Resistance R
Capacitance C_{th}	Capacitance C

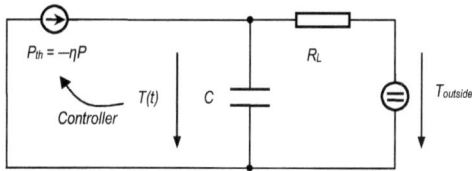

Figure 4.10: Equivalent electrical circuit model for a thermal process.

Also shown in Figure 4.10 is the link between the inside temperature $T(t)$ and the operation of the energy converter. A controller is used to operate the converter in such a way that a certain inside temperature is maintained. This is the case for types of DSM resources with thermal processes that shall be modelled with this approach. Air-conditioners control the room temperature, water heaters keep the water tank temperature and refrigerators control the inside temperature. In most cases, the controller will be a two-point regulator with

$$P_{th} = \begin{cases} P_{th,\max} \text{ for } T < T_{set} - \frac{1}{2}T_d \\ 0 \text{ for } T > T_{set} + \frac{1}{2}T_d \\ \text{previous state otherwise} \end{cases}, \qquad (35)$$

with T_{set}: set-point temperature, T_d: allowed temperature deviation amplitude.

Since the exact time behaviour of a two-point regulator is difficult to describe, the subsequent discussion will assume an ideal regulator so that $T(t) = T_{set}$ at all times.

The electrical power fed into the system is converted into thermal power non-ideally, therefore $P_{th} = -\eta P$ ($0 < \eta < 1$ is the efficiency, the minus-sign applies to cooling processes only). Now, a term for the electrical power P that is dissipated by the thermal process can be derived using $P_{th} = -\eta P$ and the circuit from Figure 4.10, which is needed for subsequent discussions:

$$P = -\frac{1}{\eta}P_{th} = -\frac{1}{\eta}\left(P_C + P_{R_L}\right) = -\frac{1}{\eta}\left(C\frac{dT_{set}}{dt} + \frac{T_{set} - T_{outside}}{R}\right) \qquad (36)$$

4.3.2 Discussion of thermal and electrical losses

Now that a generalised model for the thermal processes of interest is derived, the losses of this system can be analysed, which is important for a better understanding of how DSM measures will influence these losses. As depicted in Figure 4.11, there are two points of energy losses: first the non-ideal conversion from electrical to thermal energy, modelled by the parameter $\eta = P_{therm}/P_{el}$, and second the thermal losses due to the non-ideal isolation, modelled by the parameter R_L.

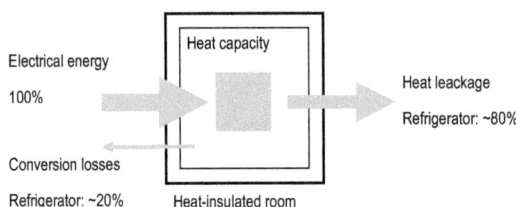

Figure 4.11: Energy flows in a thermal process (e.g. air conditioned room or refrigerator).

It is assumed that η is almost constant over the temperature range in which the system is operated. So,

$$E_{loss\ conv} = (1-\eta) \cdot E_{el}. \qquad (37)$$

During normal operation, also R_L is constant. All thermal energy that is supplied to the system will leave it via R_L. But R_L is not constant all the time. Refrigerator doors are opened; rooms are aerated; cooling chambers are refilled. All these events cause R_L of the respective system model to drop to a lower value for some time. Nevertheless, it is worth noting that these events occur independently of whether the system is used for DSM purposes or not. Consequently, stored energy lost by such events is not conceptually equal to loosing a stored DSM energy packet.

The energy losses over R_L can be calculated as

$$E_{loss,leack} = \int P_{loss}(t)dt = \int \frac{T_{outside} - T(t)}{R_L} dt \approx \frac{T_{outside} - T_{set}}{R_L} t \qquad (38)$$

(T_{set}: set-point of the temperature regulator)

Assuming that in average $T(t)$ equals the temperature set-point T_{set} of the two-point-regulator in the process, the thermal energy losses are proportional to $T_{outside}-T_{set}$.

This factually means that, when storing additional energy in the system by changing (e.g. decreasing for a cooling process) the temperature set-point, the system losses increase linear to the set-point change. However, this also means that if the set-point is changed in the opposite direction (increased for a cooling process, decreased for a heating process), the losses are linearly reduced. Thus, by taking care that the thermal process is run as long with an increased set-point as it is run with a decreased set-point, the average additional loss during DSM operation compared to normal operation can be kept zero. Due to the fact that losses are smaller than normal for lower temperature differences to the outside world, even other non-linear loss effects can be compensated by operating the system longer times with a reduced set-point. This assumes that service levels were not optimised. Although this need for compensating losses imposes additional restrictions on the overall DSM load scheduling, these can be handled fairly well as shown in the Section 4.3.4.

4.3.3 Model verification by measurements

The basic thermal model shown in Figure 4.10 has been verified using a 45 l standard household refrigerator. Measurements were conducted with an additional temperature sensor connected to an external temperature controller. The purpose of the experiment was also to analyse the reaction of the tested device on frequent interruptions of power supply. Figure 4.13 exemplarily depicts the temperature and power data for a set-point change of 3 °C.

The frequent cut from the mains supply did not cause any problems. Measurement data match the RC modelling well, although data can better be matched with a series of RC elements.

From the measurement data collected, C and R_L can be calculated. For R_L, two points with the same temperature but different times (and no set-point change in between) are considered. Since both points have the same temperature ΔT_0, the energy on the capacitor is the same at both times and has not to be considered. Therefore, all energy fed into the system between the two points gets lost via R_L.

$$\int_{t_1}^{t_2} \eta P dt = \int_{t_1}^{t_2} \frac{\Delta T_0}{R_L} dt \qquad \text{for } \Delta T(t_1) = \Delta T(t_2) = \Delta T_0 \qquad (39)$$

$$\Rightarrow R_L = \frac{1}{\eta} \frac{\int_{t_1}^{t_2} \Delta T dt}{\int_{t_1}^{t_2} P dt}$$

For determining C, the frequency f of the temperature oscillation should be used since this can be determined very precisely by counting zero-crossings within a certain time interval. By describing rising and falling edges of the temperature signal with separate exponential functions, and ΔT oscillating between ΔT_1 and ΔT_2, the thermal capacity can be calculated as follows.

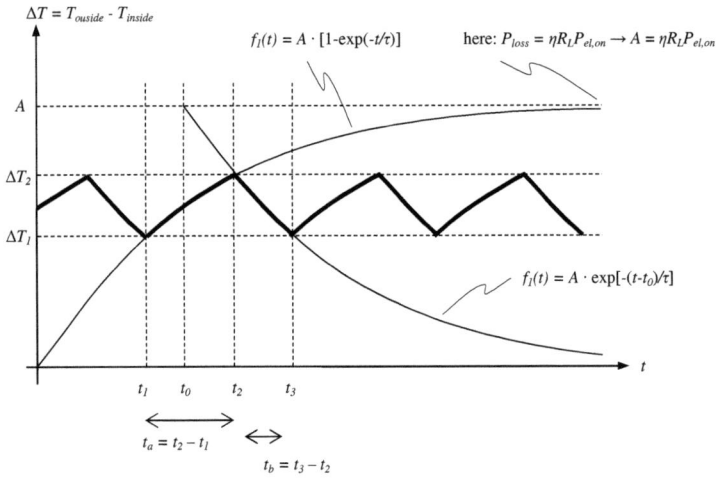

Figure 4.12: Time and temperature definitions for calculating the thermal capacity C of the observed thermostat-controlled process

Considering Figure 4.12, it can be found that $t_1 = -\tau \cdot \ln(1-\Delta T_1/A)$, $t_2 = -\tau \cdot \ln(1-\Delta T_2/A)$, $t_2 - t_0 = \tau \cdot \ln(A/\Delta T_2)$ and $t_3 - t_0 = \tau \cdot \ln(A/\Delta T_1)$. With the definitions of t_a and t_b from Figure 4.12, a term for the frequency f can be formulated:

$$f = \frac{1}{t_a + t_b} = \frac{1}{CR_L \ln\left(\frac{A - \Delta T_1}{A - \Delta T_2} \frac{\Delta T_2}{\Delta T_1}\right)} \quad \text{with } A = \eta R_L P_{el,on}. \tag{40}$$

Solving this to f, it can finally be found that

$$C = \frac{1}{fR_L \ln\left(\frac{A - \Delta T_1}{A - \Delta T_2} \frac{\Delta T_2}{\Delta T_1}\right)} \quad \text{with } A = \eta R_L P_{el,on}. \tag{41}$$

For the measurement data of the 45 l standard household refrigerator shown Figure 4.13, C and R_L can be calculated as $C = 4500$ J/K and $R_L = 0.4$ K/W.

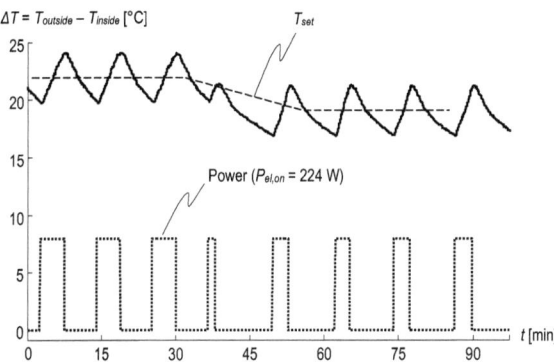

Figure 4.13: Measured temperature difference ΔT and power supply to the system for a set-point change by 3 °C (45 l Refrigerator)

4.3.4 Two options for active load shifts

Electrical appliances are commonly not designed for taking part in DSM measures. Consequently, in most cases it is not simple to interface the load to the DSM automation infrastructure. Depending on the kind of interface, i.e. whether it is possible to gain influence on the temperature set-point of the system or not, to alternatives for load shifts exist. The basic load profiles of both variants are shown in Figure 4.14. Although two-point temperature controllers are common, in the diagram a linear temperature controller is assumed for clarity.

Type A: Set-point variation

Where ever it is possible to gain access to the temperature regulator of a process, variant A "set-point variation" can be realised. Since the process temperature is proportionally increasing or decreasing with the energy level of a thermal storage (depending on whether heating or cooling is performed), charging and discharging the storage can easily achieved by varying the temperature set-point. Without DSM, the system is operated at a neutral temperature set-point $T_{set,0}$. Two other set-points $T_{set,-1}$ and $T_{set,1}$ can be found for a most applications, so that

$$E(T_{set,-1}) < E(T_{set,0}) < E(T_{set,1}), \tag{42}$$

Modelling resources for demand side management

(where $E(T)$ is the system energy level of the temperature T) and the system operation is not negatively influenced by the operation at a varied set-point temperature. The energy E_p that can be shifted by this approach is given by the heat capacity of the thermal storage and the set-point difference. While E_p is clearly restricted, the storage time t_{store} is usually not. It does not matter whether a freezer operated at -18 °C or -21 °C in terms of user comfort. Energetically, this does of course matter, but this is exactly the effect that is exploited here.

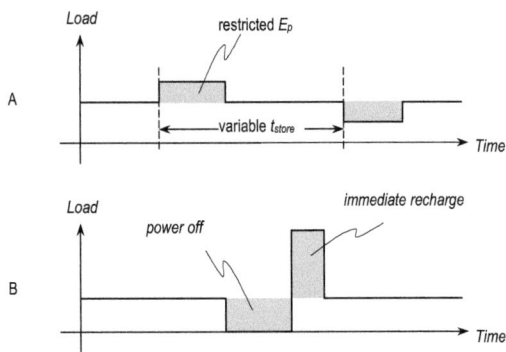

Figure 4.14: **Two different types of basic load shift functions.**
A set-point variation, B immediate recharge

Using set-point variation, storages can be pre-charged before the load reduction or post-charged after the load reduction or both. In order to achieve a maximum load reduction in a certain time interval, storages can be operated symmetrically around this interval as shown in Figure 4.15. Here, a cooling process is utilized to reduce power consumption during peak electricity demand. It is precharged in times of low demand, holds the "energy packet" to release it twice during the peak and finally post-charges after the peak. The system temperature and the system power consumption are shown, whereas the power essentially is the derivation of the temperature curve. Nevertheless, as discussed in Section 0, the temperature does also linearly influence the losses and therefore the power consumption. However, from Figure 4.15 can clearly be seen that despite this the total energy consumed by the process is the same for DSM and non-DSM operation.

The requirement for realizing the set-point variation option is, as mentioned before, that the regulator set-point setting can actually be accessed electronically. In many cases, thermal regulators are even so simple that they can easily be replaced by automatically configurable counterparts.

The advantage of influencing the regulator set-point is that the regulator itself takes care of all side effects or unpredictable events such as an opened refrigerator door. Although these events result in increased energy consumption, they would also have occurred without DSM measures and therefore can be modelled as a consumption that is independent of the storage process.

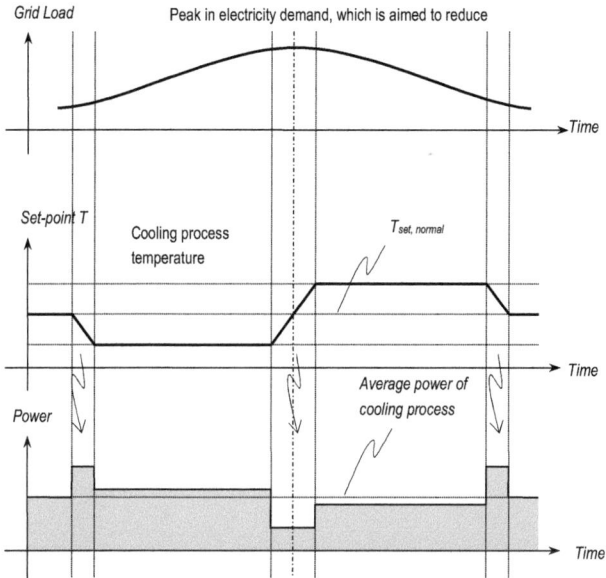

Figure 4.15: Symmetric storage management: the energy consumption of the resource with changing set-point is equal to the non-modulated version with constant set-point.

Type B: Immediate recharge

An alternative to the previously discussed storage utilisation method is the "immediate recharge" option (refer to Figure 4.14). This is the option of choice when no access to the process set-point can be gained and the only interface the process offers is its power cable. The load is simply switched off. During power-down, the energy level of the system is falling slowly (e.g. the water temperature of a boiler is decreasing). After a certain time, power is again supplied to the system. The temperature controller will immediately start to "recharge" the system, which usually takes significantly less time then the previous discharge process. There are two drawbacks of this method: First, the discharge time is determined externally. To avoid that the temperature falls to low or rises too high, a temperature sensor has to be added to the storage. Second, the storage time is restricted to the time it takes to discharge the storage. In contrast to Variant A, it is strongly depending on the thermal isolation (and isolation changes over time).

4.4 Model comparison using case study

In Section 4.2 it was shown how an example problem of DSM resource dispatch can be solved using a mathematical abstraction of the problem. There, the solution to the mixed-inter optimisation problem was not just some solution to the problem, but it is the optimal solution in a strict mathematical sense (although more than one optimal solution may exist). Nevertheless, this optimality is only valid within the simplified numerical abstraction of the problem.

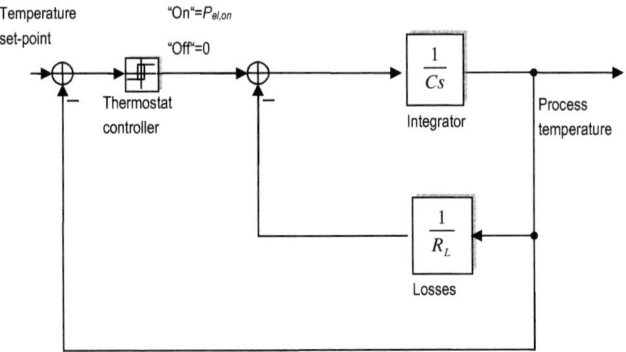

Figure 4.16: Block diagram of the simulation model

In Section 4.3, a refined model for DSM resources is described. Now, the results gained from the basic model and the refined model can be compared to each other. For this purpose, a simulation environment for thermal storages was programmed using the simulation toolset MATLAB®. The model used for the DSM resources is that presented in Figure 4.10 (page 61), using an equivalent RC circuit to represent the inert thermal process. A two-point-regulator was added, switching on and off the thermal supply according to the temperature measured at the thermal capacitance. The resulting system that was simulated is depicted in Figure 4.16. In order to achieve a representative set of DSM resources, a large number (1000) of instances derived from this basic model were simulated. R_L and C were chosen to represent an average household refrigerator. All parameters were randomly varied for each single instance within a range of +/− 20% (apart from the efficiency η that was only varied +/−5%, because otherwise unrealistic values would be reached).

The first insight gained from these simulations is that a very high numbers of such periodic power consumers are needed for achieving a reasonably smooth total load profile. The duty-cycled operation of each process causes a significant amount of distortions. Simulations with only a few individual processes have a very strong noise component. Therefore, the following simulation results are gained from simulations with 1000 individual processes, if not otherwise indicated. In Figure 4.17, the superposition of 1, 10, 100 and 1000 processes is compared. The higher the number, the better, but from 1000 simulated processes on, the achieved sum power is reasonably smooth, so that a further increase in the number of simulated processes is not necessary. (The same conclusion is drawn by Short et al. [Short07], who model the total household refrigeration load of the UK by just simulating 1000 refrigerators and scaling these up.)

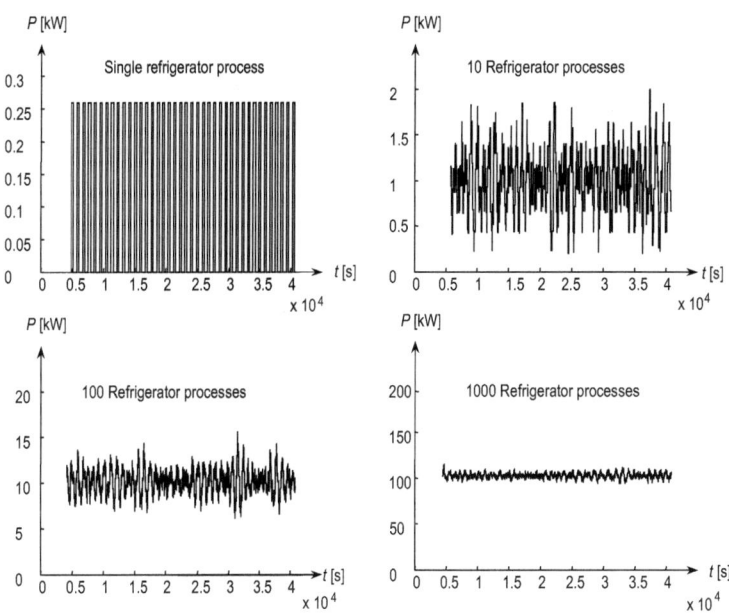

Figure 4.17: Superposition of multiple periodical loads

For demonstrating the potential lying in such a large group of relatively small DSM resources, and also for testing the simulation environment, a simultaneous set-point change of all resources shall be analysed. The set-point is changed from "normal" to "low" for 10^4 s, after that, it is set back to "normal". Definitions of "low" and "normal" are properties of each process; for cooling processes, the temperature value for "low" is actually higher than for "normal". The state names refer to the energy level of the process relatively to the outside. In average, the temperature difference between the two states "normal" and "low" is 3.6 K in this simulation. The simulation result is depicted in Figure 4.18. It can be seen that in the moment the set-point is changed, all resources pause their consumption. The total load goes down two zero, but shortly after the set-point change has occurred, processes resume to consume power. The first to resume are those processes where the process temperature was already close to the lower limit when the setpoint change occurred. The total load rises to a new level which is slightly lower than the original total load. This is due to the difference between inside and outside process temperature, which is lower for a reduced set-point, resulting in lower thermal losses. A similar reaction can be observed for the set-point increase. The resulting positive peak is shorter but has higher amplitude than the previous negative peak. This is due to the fact that the power of the chiller is higher than the losses due to thermal leakages. A refrigerator cools down faster than it warms up (under the condition that the door stays closed).

For causing two extreme peaks in the load profile, no complex scheduling is needed; a simultaneous set-point change is sufficient. This experiment was intended to verifying the simulation environment and showing some basic relations. In a second step, the two models discussed so far, the basic and the finer-grained shall be compared to each other. In order to do this, the schedule gained from a mixed-integer optimisation program (basic model) can be applied the simulation environment discussed here, which is based on the refined model.

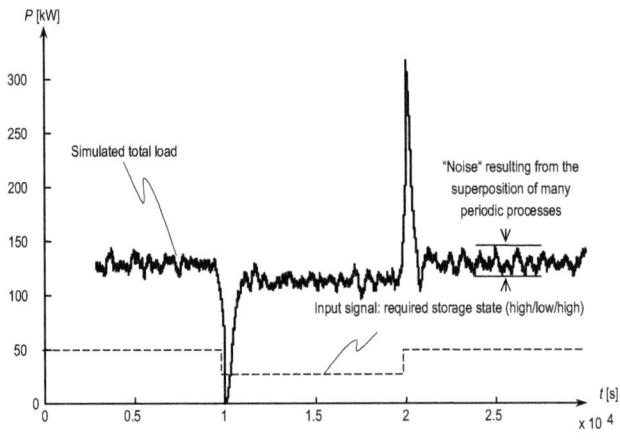

Figure 4.18: Simultaneous set-point change for 1000 DSM resources

In order to match the abstract mathematical model with parameters of more realistic electrical and thermal simulation, the actual time step and power amplitude scaling factors have to be adapted. To keep the optimisation program simple, only 50 resources are present in the optimisation and later used to schedule groups of 20 synchronised resources each rather than single resources. This results in the required 1000 simulated processes.

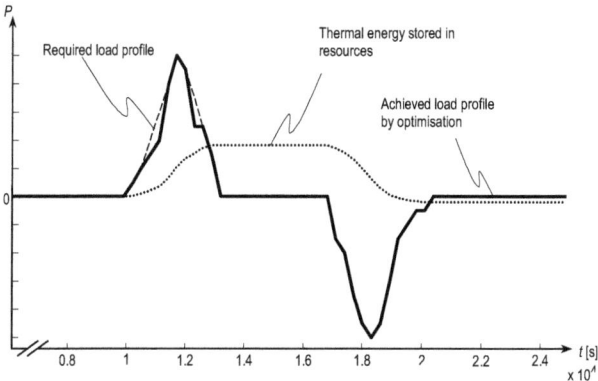

Figure 4.19: Basic model results for the basic DSM resource model. The required load profile can be followed precisely with some exceptions. Also shown is the energy stored in resources according to (27).

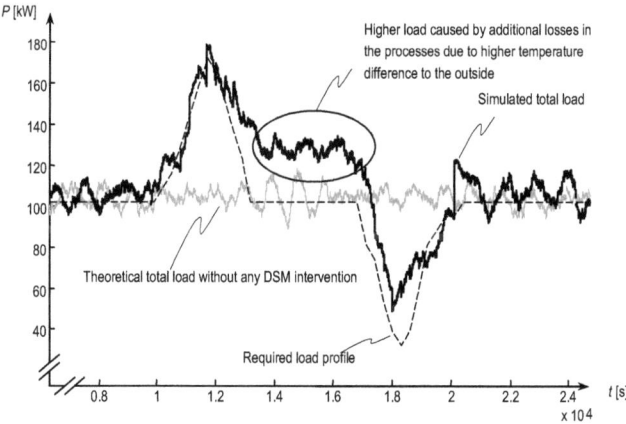

Figure 4.20: Refined model results. The schedule for load profile following results in the depicted load profile. The refined model also takes into account the individual periodic consumption patterns and the increased losses in the charged state.

Looking at Figure 4.19 and Figure 4.20, the two modelling approach can be compared. Results for the optimisation of the basic model are shown in Figure 4.19, while a fine-grain simulation result of the same schedule is depicted in Figure 4.20. The original required load profile (which is arbitrarily chosen for this example) is depicted as well as the actually achieved one in both pictures. It is clearly visible that the achieved load profile follows the required cure in both cases. Nevertheless,

71

some inaccuracies can be observed. For the basic mathematical optimisation, the only reason for inaccuracies is the small number of resources (which is only 50 in this case). If a higher number of resources had been chosen, no difference between required and achieved load profile would be visible in this scale. The predominant reason for errors in the achieved load profile of the finer-grain simulation is the noise that is caused by periodic – but because of random process parameters also stochastic – switching of the chillers, an effect that is not accounted in the basic model.

However, this noise would also occur without any dedicated DSM measures and can therefore be seen as a natural property of the loads used here. Further, it is obvious that in the middle of the simulation, when the required load profile stays at zero for a while before it becomes negative, the achieved load profile shows higher power dissipation as it was required. This is due to the fact that the load profile corresponds to a charge-discharge cycle. As mentioned before, thermal processes have higher thermal losses in the charged state. If the required load profile had been mirrored, and the charge cycle would come *after* the discharge cycle, the power dissipation would be *lower* than required.

It can be concluded that scheduling results gained from the linear optimisation can successfully be applied to finer-grain simulations of DSM resources, although the two key simplifications in the mathematical model cause drawbacks in the result accuracy. The higher the number of DSM resources in the system, the better is the accuracy. This is due to the simplification of continuous regulators in the optimisation model. Finally, the dependence of the system losses from the temperature set-point in the overall DSM system, which is not represented in the optimisation model, is significant. The refined optimisation model takes this into account.

4.5 Distributed thermal resource model

In Section 4.1 a basic DSM resource model has been outlined. The model describes the power consumed by an individual resource according to (43) as the normal resource consumption c plus the power components originating from charging or discharging the resource's energy storage.

$$p = \frac{ds}{dt} + c \tag{43}$$

In Section 4.3 the basic model has been refined for electrical loads with the capability to store thermal process energy, now taking into account storage losses. The first-order effects of the storage losses are added to (43), resulting in

$$p = \frac{ds}{dt} + bs + c. \tag{44}$$

For the equivalent electrical circuit shown in Figure 4.10, the constants b and c can be calculated by applying the definition of s. All constants are resource-specific. s refers to the energy state of the resource, specifically to the electrical energy that is (conceptually) stored. So, for thermal processes s is defined by

$$s = \frac{Q}{\eta} = \frac{TC}{\eta}. \qquad (Q\text{: Heat quantity, } T\text{: Temperature, } \eta\text{: Efficiency}) \qquad (45)$$

Consequently, using this definition and by comparing (44) to (36), it can be found that

$$b = \frac{1}{R_L C} \quad \text{and} \quad c = \frac{T_{outside} - T_{set}}{\eta R_L}. \qquad (46)$$

The outlined resource model uses a lumped resistance and a lumped capacity. The precision of the model, especially in regard to thermal processes, can be improved by assuming multiple thermal elements that have individual capacities and thermally conducting joints. Such a model is described in [Short07]. The modelled refrigerator consists of a deep-freezing part and a refrigeration part.

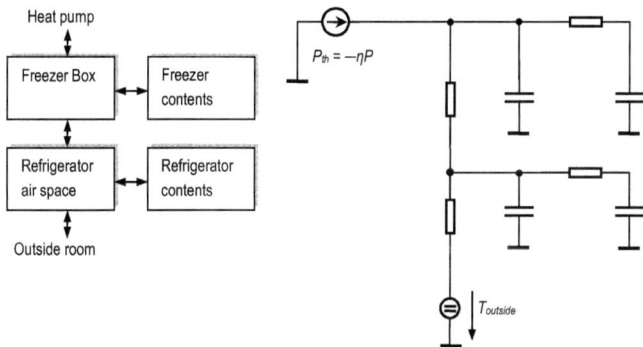

Figure 4.21: Distributed thermal model for a household refrigerator [Short07] (left) and equivalent electrical circuit (right)

This thermal model consists of the four compounds freezer box, freezer contents, refrigerator air space and refrigerator contents. In comparison to the lumped or concentrated model in Figure 4.10, the elements are now distributed. Heat flow can occur between the distributed elements as shown in Figure 4.21. Each compound has an individual thermal capacity, and the compound joints have individual thermal resistances. Figure 4.21 shows also the equivalent electrical circuit for the thermal model.

The behaviour of this distributed model for a given working point can be approximated by a lumped model. This is shown in Figure 4.22, where simulated temperature curves of the lumped and the distributed model are compared. The lumped model is empirically matched to the 4.5 °C working point of the distributed model. From Figure 4.22 it can be seen that after the set-point change to 5.5 °C, a slight mismatch between both temperature curves occurs since the lumped model is only optimised for the 4.5 °C working point.

The advantage of a distributed model is the higher precision in reproducing real-world temperature states. However, this higher precision comes with a high computational overhead. The distributed model shown in Figure 4.21 with for thermal elements requires the calculation of nine variables in each simulation step compared to a single variable for the lumped RC model. Another problem with the distributed model is that it is specialised for a certain kind of DSM resource. The given example models only a household refrigerator. For any other thermal resource, e.g. an air-conditioning system or a water boiler, the distributed model is different.

Due to the fact that the distributed model is less flexible, requires much more computation while improving the model precision only slightly, the lumped RSC-model should be preferred.

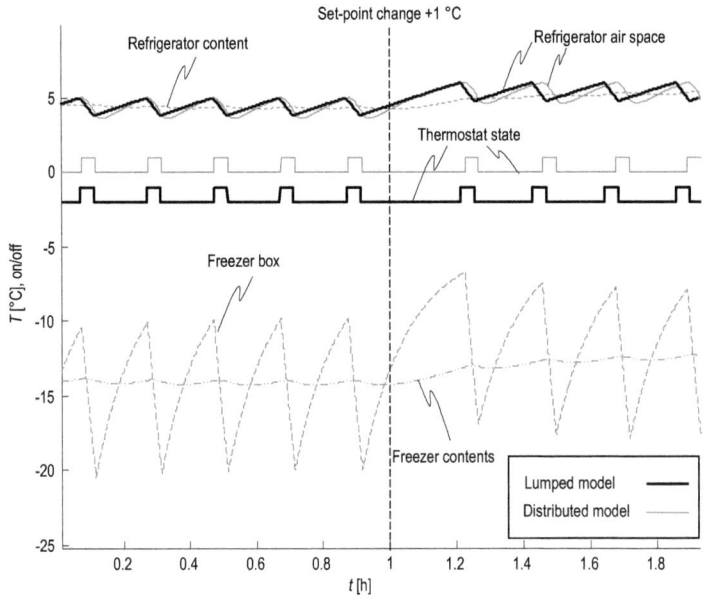

Figure 4.22: Simulated temperatures of a household refrigerator, comparison of lumped and distributed model.

5. Primary control algorithm using distributed resources

The model for DSM resources derived so far can be interpreted as follows: Provided adequate process interfaces infrastructure (access to set-points) and communication channels (for coordinating load shifts) are in place, the capability of energy storage in a large group of loads can be utilised without interfering with the safe and uninterrupted operation of these loads.

The *service* of energy storage on the demand side has to fit into an existing or future market construct, which enables to gain profit from such a system. Within the current market framework, primary control is the most attractive application (see discussion in Section 3.1). Control power is only supplied for a few minutes, until primary control is substituted with secondary control measures. Further, primary control power provision is depending on the system frequency, which can be used as an implicit communication channel.

5.1 Deduction of the algorithm

The provision of primary control power is determined by the system frequency f, which can be measured at every point of the grid. The grid frequency therefore acts as an implicit, unidirectional, real-time and highly reliable communication channel. A system of DSM loads that takes part in primary control has to adhere to a given power characteristic depending in the frequency. The relationship of the power that has to be provided and the network frequency f can be expressed in a function $K(f)$, subsequently referred to as power function. A basic system that fulfils the primary control objective is shown in Figure 5.1. The load or generation of a single unit connected to the power system is modulated with $K(f)$. This setup becomes part of the overall primary control system as discussed in Chapter 1.4. The term modulation is used to point out the fact that the load (or generator) behaviour is only changed *relatively* to its usual behaviour without primary control. Therefore, the delta character (Δ) will be added to variables to reflect this (e.g. ΔP in Figure 5.1).

The speciality of the algorithm that shall be discussed here is that it is not operating on a single unit but rather coordinating a large number of potentially small contributors to primary control. However, before dealing with the aspects of distributed contributors, the power function $K(f)$ shall be discussed more in detail since it is the main objective of the system to realise this function as precisely as possible.

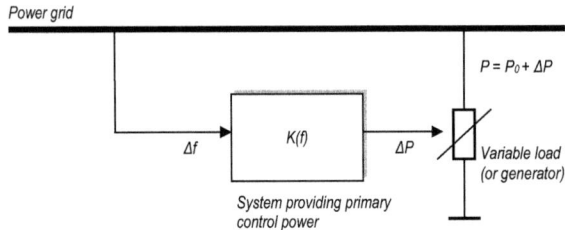

Figure 5.1: Block diagram of a system providing primary control power. The power is modulated around a working point P_0 which reflects the normal operation without primary control.

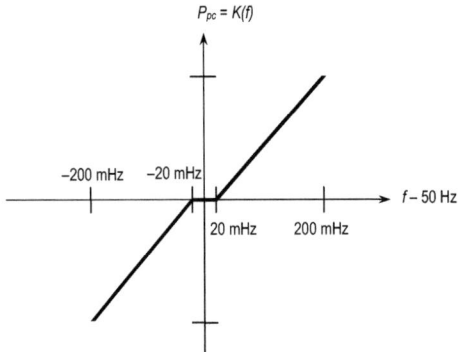

Figure 5.2: Primary control power as function of the network frequency $K(f)$. Outside the dead band of +/- 20 mHz around the nominal value of 50 Hz the provided control power grows linearly with the frequency deviation.

As stated in the ground rules of the Union for the Co-ordination of Transmission of Electricity (UCTE) [UCTE04a, UCTE04b], the provision of primary control power is linearly increasing with the system frequency deviation from 50 Hz. In order to avoid too frequent control activity, a dead band of +/- 20 mHz is introduced [UCTE04a]. As long as the system frequency is within this dead band, no control activities are issued. Only when the frequency exceeds the dead band, primary control power is supplied. The resulting frequency-power function is shown in Figure 5.2. It is essentially equal to the droop characteristic discussed in Section 2.1.1, but there are two differences: first, the 20 mHz dead band is reflected, and the whole curve is mirrored compared to the droop curve, since it describes the frequency response of loads and not of generators.

From Figure 5.2 it can be seen that the power function is piecewise linear and monotonously increasing. These two properties will be used by the algorithm. In fact, the algorithm can work with any function $K(f)$ that has these properties. $K(f)$ determines the power for primary control P_{pc} of a system taking part in the primary control measure. Such a system is characterised by slope of the $K(f)$ function, which determines how intense the frequency response is.

The differential power ΔP consumed by a system of DSM loads, which is defined relatively to the normal resource operation power, is the sum of all individual differential power amplitudes Δp_i. The system is linear in this regard. Now, an algorithm is searched that influences the individual resources in such a way that their total differential power ΔP equals (to a certain extend) the required power P_{pc}. The relationship of ΔP, Δp_i, P_{pc} and $K(f)$ is shown in (47).

$$\Delta P = \sum_{\forall i} \Delta p_i \rightarrow P_{pc} = K(f) \tag{47}$$

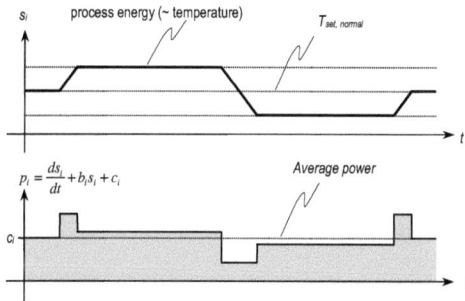

Figure 5.3: Energy level s_i and consumed power p_i of a DSM resource i according to (48). When the storage energy level is increased, the resource consumes additional power. When the energy level is decreased, the resource consumes less power.

So far, all considerations were independent of the DSM resource model. Now, in order to be able to find and expression for the individual power Δp_i, the energy storage model discussed in Chapter 4 will be facilitated. For the design of the distributed control algorithm, a linear modelling is of great advantage since this allows the definition of some system-wide sum variables (such as ΔP), that can easily be broken down to individual resources. In Section 4.3.4, the basic possibilities of influencing loads have been discussed. It was found that the simply (because direct) control of the process power p_i by switching the load off and on is not the preferable solution since it disregards internal process knowledge and has only short-term effects. The more elegant and better solution is to gain influence on process variables that only influence the power indirectly. For thermal applications, this is the process temperature set-point $T_{set,i}$. However, it is possible to generalise this using the energy storage model as outlined in Section 4.1.1. Each storage i (either conceptual or real) has an internal energy state s_i that indicates the energy currently stored in the storage. It has already been discussed that the power consumption of the DSM resource can be described as the power that is needed for compensating the losses of the process, superposed by the power consumed (or released) by charging/discharging the internal energy storage. This is depicted in Figure 5.3. When the storage energy level is increased, the resource consumes additional power (compared to the normal or average operation). When the energy level is decreased, the resource consumes less power. Additionally, in most cases the charged storage has higher losses than the discharged.

This basic behaviour of a DSM resource can be summarised as in (48) and (49). In (49), the average power is omitted and only the differential power is described:

$$p_i = \frac{ds_i}{dt} + b_i s_i + c_i \tag{48}$$

$$\Delta p_i = p_i - c_i = \frac{ds_i}{dt} + b_i s_i \tag{49}$$

where Δp_i is the differential power consumed by resource i, s_i is the energy stored in resource i and b_i, c_i are a process-specific parameters. For an ideal storage without losses, $b_i = 0$.

As discussed in Section 4.3.4, s_i is the process parameter that can be externally controlled. The maximum amount of energy that can be stored in the resource is determined by the value range of s_i. The value ranges of s_i as well as the deviation of s_i are bounded within certain limits, which are process-specific.

5.1.1 Discussion of the simplified approach with $b_i = 0$

In the following, it will first be assumed that all $b_i = 0$, i.e. all storage processes in the distributed resources are ideal and no additional losses are caused. This simplifies the problem and allows describing the general idea of the approach used here. Later, this simplification will be dropped and the case $b_i > 0$ will be discussed.

Since the DSM resource model is linear, a system-wide storage energy S can be defined:

$$S = \sum_{\forall i} s_i \tag{50}$$

This definition is motivated by the search for a relationship between the system frequency f and the individual process energy s_i. For $b_i = 0$, (48), (49) and (50) can be combined to (51):

$$\Delta P = \frac{dS}{dt} = \sum_{\forall i} \frac{ds_i}{dt} = K(f(t)) \tag{51}$$

By integration of (51), the following is gained for the required system-wide energy state S:

$$S = \int K(f(t))dt \quad \text{(note: } K(x) \text{ is not a function of time, see Figure 5.2)} \tag{52}$$

The objective for a collective of distributed DSM resources taking part in primary control is to adhere to the power function $K(f)$. The individual resource has a restricted information horizon. It is not aware of how much power the other resources consume, or what their energy states are. It only can control its own energy state s_i, therefore also its own power consumption, and it can measure the network frequency f. Now, (52) gives a calculation directive how to calculate the system-wide energy state S only from the system frequency. So, S is actually known at all places and the remaining question is how to determine the individual storage state s_i from S at the place of resource i without any knowledge about all other storage states s_j.

Two basic solutions can be applied here. The first, which shall be called the "continuous approach", simply scales down S to the individual s_i, so that

$$s_i = k_i \cdot S \quad \text{with} \quad \sum_{\forall i} k_i = 1 \tag{53}$$

While it is simple and straight-forward, the drawback of this solution is that s_i continuously follows S, resulting in a very fine-grain variation of the resource's energy state. Any small error (e.g. due to quantisation) is multiplied with the potentially very large number of resources and can have major impact on the system performance. Further, all resources would be operated in a synchronous fashion, potentially causing catastrophic resonances among periodic loads such as two-point regulated thermal processes.

The second option, the "discrete approach", allows only discrete values for s_i:

$$s_i = r_i \cdot s_{i,\max} \quad \text{with} \quad r_i \in \{-1; 0; +1\} \tag{54}$$

The state 0 refers to a normal, unchanged set-point or resource energy level. In state +1, the system is positively charged, having an increased energy level compared to state 0, whereas in state −1 the resource has a lower energy level compared to state 0. By choosing this approach, both disadvantages of the continuous approach are avoided. All resources commit either no or their maximum storage potential to the system; and changes between these states are non-synchronous between all resources.

In order to determine when s_i changes its state, the positive value range of S is divided into n sub-intervals of size $s_{i,\max}$, resulting in a set of interval borders L = $\{l_0, l_2, \dots l_{n-1}\}$. This can also be done to the negative half of the value range without any differences since both halves are symmetrical. The current state of r_i is determined by (55).

$$r_i = \begin{cases} +1 \text{ for } S > l_i \\ -1 \text{ for } S < -l_i \\ 0 \text{ otherwise} \end{cases} \quad \text{and} \quad s_i = r_i \cdot s_{i,\max} \tag{55}$$

In Figure 5.4 an example is shown how the individual s_i are chosen according to the development of S over time. First off all, it can be seen that S is calculated by integrating over $K(f)$. When S exceeds the first activation level l_0, which is zero because it is the first one, the first resource s_0 is activated. After a while, S exceeds the next activation level and s_1 is activated. This goes on until S is no longer increasing. In this example, only six resources become activated. In reality this will usually be many more.

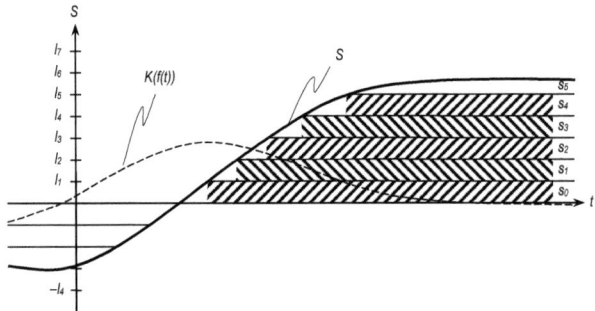

Figure 5.4: The total value of S is constituted by multiple individual resources that commit s_i each. Each individual resource i can calculate S and knows its activation level l_i. When S exceeds this level, the resource becomes active and changes its own energy state s_i accordingly.

Considering Figure 5.4, it becomes intuitively clear that Resource 0 will be active more often than Resource 5. This is due to the non-homogenous probability density function of S. The probability density function of S plays a strong role for fair workload distribution in the system and will be studied more in detail at a later point (Section 5.3). At this point it should only be noted that without additional measures, the workload will be distributed unevenly over resources. The most straightforward solution for this is to re-arrange the activation levels over the value range of S from time to time so that heavily used resources will be re-situated in less frequently activated intervals. While the system was designed up to this point for only relying on the system frequency measurement as implicit means of communication, the task of re-arranging the activation levels requires a dedicated communication infrastructure.

5.1.2 Structure of the resulting system

Taking into account the structural considerations from the previous sections, the following structure for the system of distributed DSM resources providing primary control power is proposed, which is also depicted in Figure 5.5 (whole system) and Figure 5.6 (single resource). A set of resources is connected to the power network. Each resource has the capability for storage of the energy $\pm s_{i,\max}$

and its energy state s_i can be changed at any time (compare Type A in Section 4.3.4). All resources measure the system frequency f and use this to calculate the required system-wide energy state S. It can be assumed that the value of S is precisely synchronised and equal in all resources. Technical measures to achieve this are subject to discussion in subsequent sections.

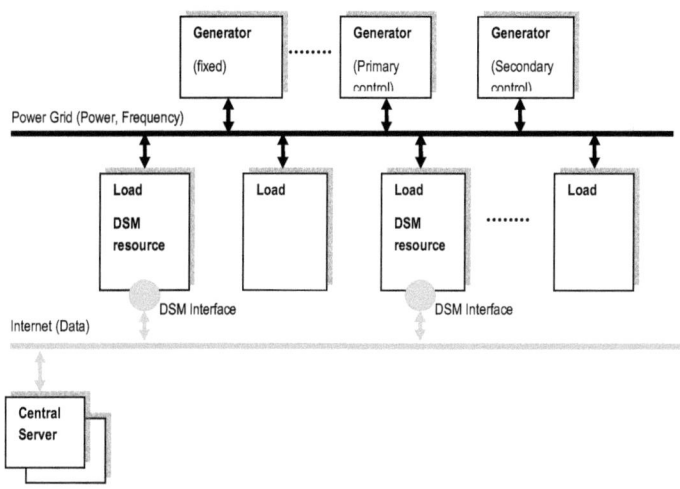

Figure 5.5: Proposed system with primary control provided by generators *and* a DSM system consisting of multiple participants coupled by the system frequency f and a supporting communication network

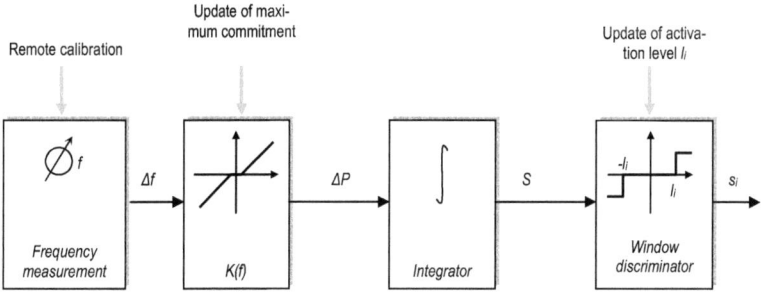

Figure 5.6: IRON-CPP algorithm in the DSM Interface. Local set-points s_i are determined on the basis of frequency measurement. Upper arrows indicate non-real-time updates via the communication infrastructure.

Each resource measures f, calculates $K(f)$ and integrates the result over time. Then, it uses its individual activation level l_i to determine its energy state. Changes in the energy state will result in a change of power consumption of the resource, and the system is designed in such a way that all individual changes in power consumption add up to realise the required power characteristic given by $K(f)$. The proposed algorithm is subsequently called "IRON-CPP" (Integral Resource Optimisation Network – Control Power Provision).

The IRON-CPP algorithm is executed on each single resource and uses the system frequency as only input. Therefore, a fast and dependable reaction on frequency changes is guaranteed. Nevertheless, there is a need for an additional communication infrastructure that serves for multiple purposes as outlined in Figure 5.6 and listed below.

- The precise measurement of the system frequency is crucial for a well-synchronised calculation of S. Since the system frequency is the same in the whole power system, it can be measured at any place. Measurement results at the resource's site can be compared to a high-precision central measurement and a remote calibration of the local measurement units can be performed.

- The number of resources taking part in the system can vary. Therefore, also the committed maximum power for primary control can change. Since this information is part of the power function $K(f)$, it must be possible to update the versions of $K(f)$ stored at individual resources.

- As discussed before, the activation levels have to be re-arranged over the value range of S on a regular basis for maintaining a fair workload distribution in the resources.

However, in contrast to the measurement of f, the information exchange over the dedicated communication channel is not time-critical. Communication can be realised in a peer-to-peer-fashion between the resources, or using a central server architecture as suggested in Figure 5.5. This issue is discussed in Section 5.3.

5.1.3 Improved algorithm for the case $b_i > 0$

So far it was assumed that the resource behaviour (power consumption) is only depending on the deviation of s_i ($b_i = 0$ in (49)). In reality though, there are also linear and even non-linear power components. Since this algorithm for primary control is based on the linear resource model from Chapter 4, non-linear components will be assumed to be so small that they can be neglected. Nevertheless, the linear components of the resource power cannot be neglected, which can be concluded from an example shown in Figure 5.7. In this figure, the system power for ideal resources ($b_i = 0$) and real resources ($b_i > 0$) is compared. The shown trajectories have been calculated on the basis of real frequency data, using (49) as continuous resource model. For the case $b_i > 0$, a realistic value for b has been assumed that was calculated for a household refrigerator ($b = {}^1/_{1440s}$). The result shows qualitatively that neglecting the linear power component is not possible. Even for relatively small losses in the process, the trajectory is far off the required power function $K(f)$. A solution for overcoming this major obstacle can be found, however this solution shall be motivated by another problem of the current system design.

In the previous discussion the value range of the overall system energy S plays a role. The basic idea of the system is to choose S in such a way that the deviation of S, which is essentially the system power ΔP, meets exactly the requirements of the power function $K(f)$. This is done by integrating $K(f)$ over time. So, the value range of S is actually the value range of the integral

$$S = \int K(f(t))dt.$$ (56)

Theoretically, the system frequency $f(t)$ has a perfectly symmetrical probability density. So, it should be guaranteed that the value of the integral crosses zero from time to time and that the integral is not divergent. However, this does not give any guaranties concerning the bounds of S. In fact, even if the probability density function of $f(t)$ is perfectly symmetrical, the slightest measurement offset will cause the integral to become divergent.

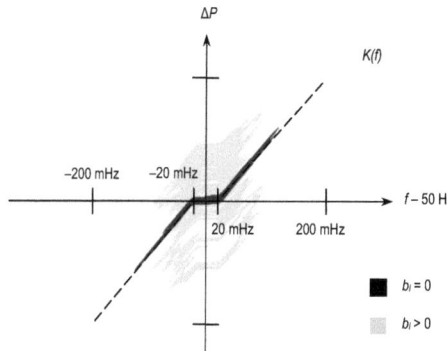

Figure 5.7: Example trajectories of the overall differential system power ΔP. For $b_i = 0$ the simplified resource model fits well and results in dynamic operation come close to the required $K(f)$. However, with realistic values for b_i the simplified resource model is no longer sufficient.

A pragmatic but also efficient way of avoiding the value of the integral to exceed all bounds is to migrate from an ideal integrator to an integrator with "losses". This can also be seen as a low pass filter that is transparent for short-term changes (higher frequencies) but filters out the long-term increase or decrease of the integral value (lower frequencies). In an electrical circuit, a capacitor in combination with a resistor would be used, so that the loss in capacitor voltage is proportional to the voltage itself, resulting in an exponential decrease. The ideal integrator can be exchanged by such a component in the hope that the impact of this replacement on the system operation is only minor.

In order to calculate the impact of a non-ideal integrator, the system shall be examined in the frequency domain. Before doing so, some general matters and conventions should be considered:

- The system is seen as a continuous system and therefore the continuous Fourier transformation is applied according to [Ohm06, p. 29]. In practice however, the system will probably be implemented in a time-discrete fashion. The findings of the continuous analysis can also be derived using time-discrete analysis methods such as z-transformation [Ohm06, p 95].

- In order to avoid the system frequency being mixed up with the frequency domain variable f, the application of the power function on the system frequency $K(f(t))$ will be treated as basic signal $K(t)$ in the time domain and $K_f(f)$ in the frequency domain.
- In general all signals in the time domain will now be written with t as argument (e.g. $S \rightarrow S(t)$, $s_i \rightarrow s_i(t)$) and in the frequency domain with f as argument and subscript (e.g. $S_f(f)$).

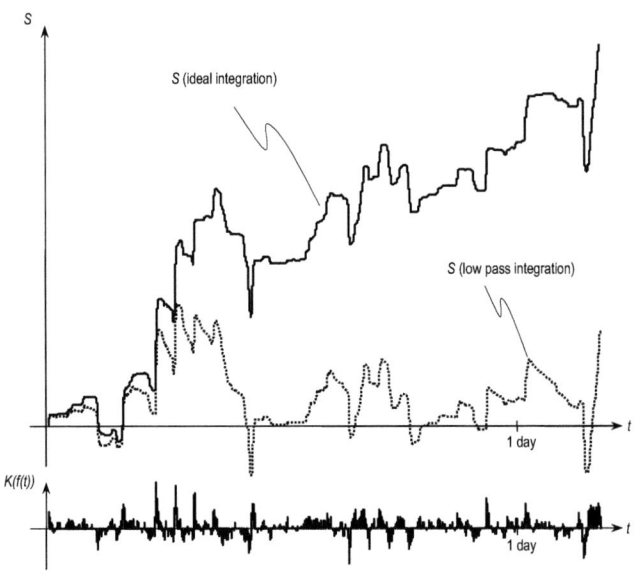

Figure 5.8: The time-domain comparison of ideally integrated and low pass integrated frequency data measured over 28 hours reveals that the diverging trend of the ideal integration can be successfully avoided.

At first, the system with ideal integrator shall be considered. By transforming (56) into the frequency domain, the following is achieved:

$$S_f(f) = \frac{K_f(f)}{j2\pi f} + \underbrace{\frac{1}{2}K_f(0)\delta(f)}_{\text{constant component } K_f(0)=0} \tag{57}$$

The second term falls away since the constant component of the frequency signal is zero. For achieving the total system power, (49) is used and it is assumed that

$$b_i = b \, \forall i \tag{58}$$

in order to be able to give a closed form of the solution:

$$\Delta P_f(f) = \underbrace{K_f(f)}_{\text{intended solution}} + \underbrace{b\frac{K_f(f)}{j2\pi f}}_{\text{error term}} \qquad (59)$$

Here, the second term transforms to the integral over $K(t)$ in the time domain, which is an error term that leads to the strong discrepancies between intended and achieved power trajectory as shown in Figure 5.7. However, for $b = 0$ the result equals the intended solution.

In a second step, the non-ideal integrator shall be considered. While the ideal integrator has the impulse answer $\varepsilon(t)$, now a system with the impulse answer $\varepsilon(t)e^{-t/\tau}$ is chosen, where τ is a parameter that determines how fast the impulse answer approaches zero. Both impulse answers are depicted in Figure 5.9. In the time domain, this application of a different kind of integrator results in a new version of (56):

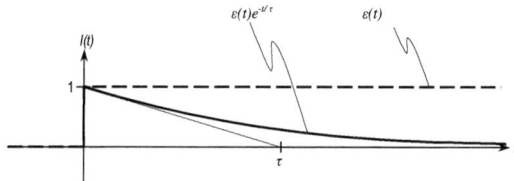

Figure 5.9: Impulse answers of ideal integrator $\varepsilon(t)$ and low pass integrator $\varepsilon(t)e^{-t/\tau}$ in the time domain.

$$S(t) = K(f(t)) * \left(\varepsilon(t)e^{-\frac{t}{\tau}}\right) \qquad (60)$$

Here, the operator * stands for the continuous convolution, which is defined by [Ohm06, p. 13] as

$$a(t) * b(t) = \int_{t'=-\infty}^{+\infty} a(t)b(t'-t)dt' . \qquad (61)$$

Equation (60) can again be transformed into the frequency domain,

$$S_f(f) = K_f(f)\frac{\tau}{1 + j2\pi f\tau} , \qquad (62)$$

and again (49) can be used to achieve a term for the total system power:

$$\Delta P_f(f) = K_f(f)\frac{j2\pi f}{1 + j2\pi f} + bK_f(f)\frac{\tau}{1 + j2\pi f\tau}$$

$$= K_f(f)\underbrace{\frac{j2\pi f + b\tau}{1 + j2\pi f\tau}}_{=1 \text{ for } \tau = \frac{1}{b}} = \underbrace{b\tau}_{=1 \text{ for } \tau = \frac{1}{b}} K_f(f)\underbrace{\frac{1 + j2\pi \frac{1}{b}f}{1 + j2\pi f\tau}}_{=1 \text{ for } \tau = \frac{1}{b}} \xrightarrow{\tau \to \frac{1}{b}} K_f(f) \qquad (63)$$

It can be seen from (63) that if τ is chosen to equal b^{-1}, then all additional terms fall apart and the system power is exactly equal to the required power function $K_f(f)$. This is a positive surprise, since it was expected that the non-ideal integrator motivated by the divergence problem would have negative impact on the result. But obviously the negative effects of non-ideal integrator on one hand and the linear power component on the other hand compensate each other ideally for $\tau = b^{-1}$. This can also qualitatively be explained: the linear power component modelled by the parameter b is the result of non-ideal energy storage. Losses are proportional to the absolute value of energy stored in the resource. Now, in the algorithm to determine the required individual storage state s_i for a given power amplitude, this behaviour reproduced by the non-ideal integrator.

Still, the question remains whether τ can actually be chosen to equal b^{-1}. Since τ is a free parameter of the non-ideal integrator, any value can be assigned to it. But b is just a simplification of the set of individual resource properties b_i (see (58)). Each resource i has its individual value of b_i depending on the quality of its storage isolation. Only for perfect lossless storage, $b_i = 0$. However, τ is a system-wide parameter that cannot be adapted to individual resources since it is used to calculate $S(t)$ which has to be the same in all resources for a precise synchronisation of charging and discharging activities. Consequently, $\Delta P(t) = K(f(t))$ cannot be exactly achieved, rather a good compromise for τ has to be found. Nevertheless, it has been shown by (63) that it is possible by careful selection of resources and parameters to come arbitrarily close to the requirement $\Delta P(t) = K(f(t))$. The actual selection of τ is discussed in Section 6.1.5, only after a method of measuring the performance of the algorithm in simulations has been presented.

5.1.4 Time-discrete realisation

The calculation of S is given by (60) for the time-continuous case. However, for a practical implementation of the proposed IRON-CPP algorithm, and also for simulations, a time-discrete rule for calculating S is needed. This can be obtained as follows:

Starting point is the frequency-domain definition of S, as given in (62):

$$S_f(f) = K_f(f) \frac{\tau}{1 + j2\pi\tau f} \tag{64}$$

Multiplication of both sides with $1 + j2\pi\tau f$ results in

$$S_f(f) + j2\pi\tau f S_f(f) = \tau K_f(f). \tag{65}$$

This equation can be transformed back into the time domain using the fact that the multiplication with $j2\pi f$ actually denotes a derivation:

$$S(t) + \tau \frac{dS(t)}{dt} = \tau K(f(t)). \tag{66}$$

In the next step, the deviation can be replaced by a differential quotient. By omitting the limes and exchanging dt into Δt, the equation is now time-discrete.

$$\frac{S(t+\Delta t)-S(t)}{\Delta t} = K(f(t)) - \frac{1}{\tau}S(t) \qquad (67)$$

Finally, the equation can be re-arranged so that $S(t + \Delta t)$ is on the left side:

$$S(t+\Delta t) = (K(f(t))-\frac{1}{\tau}S(t))\Delta t + S(t) = \Delta t K(f(t)) + S(t)\left(1-\frac{\Delta t}{\tau}\right). \qquad (68)$$

This equation expresses how to calculate S in the next time step based on the time step width Δt and the current value of S.

5.2 Grid frequency

The grid frequency, or more precisely the changes over time of the grid frequency, is the input signal to the algorithm described above. The IRON-CPP algorithm essentially performs an integration of the frequency deviation from the nominal value and charges/discharges distributed storages on the basis of this integral signal.

An example for the grid frequency behaviour is shown in Figure 5.10. In the observed time period the deviation was never larger than 85 and never smaller than –65 mHz. It usually exceeds the dead band of +/– 20 mHz only for a few minutes in one direction, then returning to zero and exceeding into the other direction.

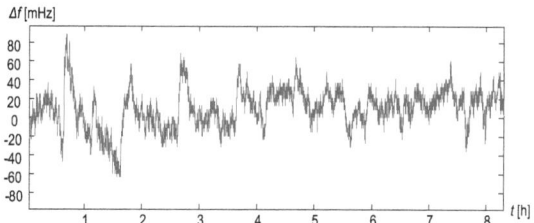

Figure 5.10: Example for the grid frequency deviation from the 50 Hz nominal value over time (source: own measurements, starting at 01.03.2007, 14:31:20, local time Vienna, Austria)

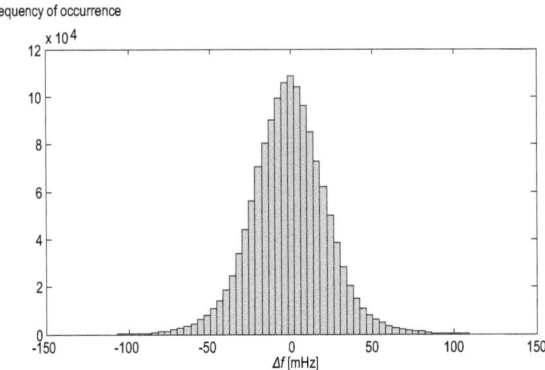

**Figure 5.11: Histogram of frequency deviation data measured over three weeks
(source: own measurements)**

In practical power grid operation, the frequency deviation is kept in a range of +/− 100 mHz. This can be seen from the histogram shown in Figure 5.11. The distribution of the frequency deviation values is the result of regularly occurring imbalance situations in the grid and it reflects the error signal of the power-frequency-control (see Section 2.1). Nevertheless, the distribution is close to a Gaussian distribution, a fact that forms the basis of the performance metric that will be discussed in Section 6.1.2.

The deviation mean value is always 0. The reason for that is that additionally to primary, secondary and tertiary control, there is also a mechanism of *time control* [UCTE04b]. The grid frequency is used as a time basis for many clocks connected to (and powered by) the grid. For this time basis to be of adequate accuracy, the time control mechanism keeps the frequency average at exactly 50 Hz [UCTE04b]. However, it cannot be assumed that the frequency mean value of actually measured time series is exactly 50 Hz, since inaccuracies in the measurement equipment can occur. Nevertheless, the 50 Hz mean value could be used for calibrating the measurement equipment.

5.3 Workload balancing by activation level update

The IRON-CPP algorithm provides primary control energy to the power grid by controlling electrical loads. As discussed above, the controlled loads are modelled as energy storages that can be charged and discharged. In fact, the energy storages (or DSM resources) have three states: neutral, positively charged and negatively charged (compare (54)). The power consumption of the resource is primarily determined by the state changes (Figure 2.1). IRON-CPP issues state changes in the order of activation levels associated with the resources as described in (55). However, due to the order of resource activation, resources with lower numbers will always be activated more often than resources with higher numbers. Without additional measures, the workload will be distributed unevenly over resources as depicted in Figure 2.1. The following sections will discuss this problem and potential solutions more in detail.

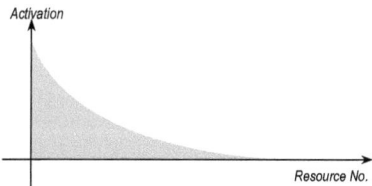

Figure 5.12: Uneven distribution of activation times over resources in case of no activation level update. Resources with lower numbers are activated more frequently than resources with higher numbers.

First of all, it should be more clearly defined what "resource activation" means. A resource is energy storage in the sense of Chapter 4, i.e. an electrical load that can be used for load shifting. Activation means that the storage state of the resource is changed to a non-neutral value, so it is either negatively or positively charged. The activation ends once the resource state is switched back to neutral. In practice, the state of a resource can e.g. refer to the set-point temperature of a heating or cooling process. Then, the neutral state means that the set-point is unchanged and reflects exactly the users or owners wish, and charged means an either slightly increased or decreased set-point. In general it is reasonable to assume that the preferred state of the resource is "neutral", i.e. no change to normal operation. So, the activation workload in the system should be evenly distributed over all resources.

The relationship between resources, activation levels and resource activation is depicted in Figure 5.13 for clarification. As described in Section 5.1, at time t all resources with activation levels below the frequency integral function $S(t)$ are activated. The most straight-forward solution for a fair workload distribution is to re-arrange the activation levels from time to time so that heavily used resources will be re-situated on the activation level scale to less frequently activated intervals. The task of re-arranging the activation levels requires a dedicated communication infrastructure. However, this infrastructure has only to fulfil moderate demands in terms of delay, bandwidth and availability. Since the activation level update has only to be performed in longer time intervals (once in several days), only very few data has to be exchanged and the actual change can be planned in advance so that there is no on-line connection necessary at the time of update. (This assumes that all nodes have synchronised clocks.)

Primary control algorithm using distributed resources

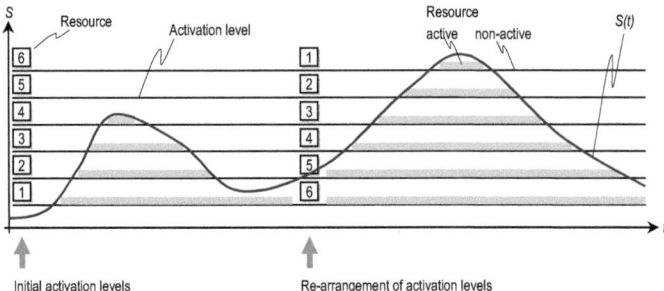

Figure 5.13: Arrangement of resources and activation levels. Each resource is assigned to an activation level. All resources on levels below S(t) are activated. This assignment has to be changed regularly to achieve a fair resource allocation.

For the re-arrangement of resources, a number of possible algorithms exist. Many process scheduling algorithms from the domain of operating systems can be adapted to serve for the given application here. Two different approaches have been selected for a more detailed discussion and comparison. One is the "random technique", which simply permutates the activation levels randomly. It is a straightforward and algorithmically simple approach, i.e. it does not require much memory or other computational resources. The second is the "feedback technique" [Stal05], which is especially suited for scheduling problems where the number of activations is not known *a priori*, which is the case here. The feedback technique uses observations on the past activation history for determining the next resource to be activated.

5.3.1 Simulation and comparison of permutation algorithms

The performance of the two approaches Random Permutation and Feedback Permutation is evaluated in a case study simulation. In this simulation, a set of $N = 1000$ resources is considered. Each resource has an individual energy commitment $E_{shift}(i)$ ($i = 0...N-1$), that is calculated using a base value ($\max(S(t))/1000$) and a +/-50 % randomisation. Initially, the activation levels l_i are incrementally calculated in the order of the resource number i as outlined in (69).

$$l_i = \sum_{k=0}^{i-1} E_{shift}(i) \tag{69}$$

The function $S(t)$ is calculated according to (68) on the basis of a 24 h measurement of the grid frequency (Figure 5.14). This measurement is regarded to be representative, since it results in a shape for the S function with considerable dynamics, while the dynamics of the S function are restricted by the non-ideal integration. Using this $S(t)$, the resource activation is simulated by adding up the number of simulation time steps in which the particular resource was activated ($l_i < |S(t)|$). The result is divided by the number of simulation time steps. Thus, a resource activation of 1 means that the resource was active for the duration of the complete observation period, a value of 0 denotes that it has not been active at all.

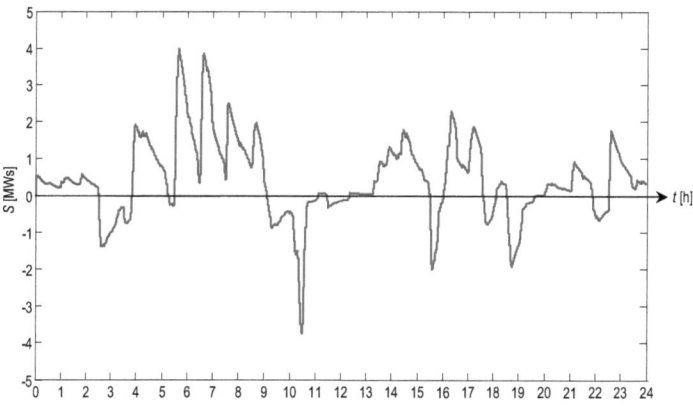

Figure 5.14: The function $S(t)$ calculated according to (68) on the basis of a 24 h measurement of the grid frequency.

During this simulation, three different permutation approaches are implemented:

1. **No Permutation.** This is used as a reference scenario to show the unfair resource allocation in case of no permutation.

2. **Random Permutation.** Here, in certain time intervals Δt_p the activation levels are re-calculated, but not in the order of resource numbers as in the initial step but on the basis of a random resource order. This is done by obtaining a random permutation $p(i)$ of the resources and then calculating the activation levels as outlined in (70).

$$l_i = \sum_{k=0}^{i-1} E_{shift}(p(i)) \qquad (70)$$

3. **Feedback Permutation.** For the Feedback Permutation, the activation levels are also re-calculated in certain time intervals Δt_p. This time, the order for the incremental addition is determined by the number of activations in the activation history. Those resources with lower activation numbers are first.

The simulation results are obtained for a permutation interval of $\Delta t_p = 0.5$ h (resulting in 48 permutations during the observation period) and are depicted in Figure 5.15. In a realistic application, the permutation interval can be chosen much longer, since the observation interval is also considerably more extended. The workload fairness needs only to be achieved over weeks of months. In Figure 5.15, the resource activation is shown over the resource number. The curve for "No Permutation" resembles the shape that has already been anticipated in Figure 5.12. "Random Permutation" improves the situation considerably, but still a strong deviation in the activation of different resources can be observed. Finally, with "Feedback Permutation" the best result is obtained. The deviation of the latter two result curves is depending on the permutation interval of Δt_p. The more often the permutation happens in the observation period, the better is the result. For achieving similar results with Random Permutation and Feedback Permutation, Random Permutation needs to be issued more often than Feedback Permutation.

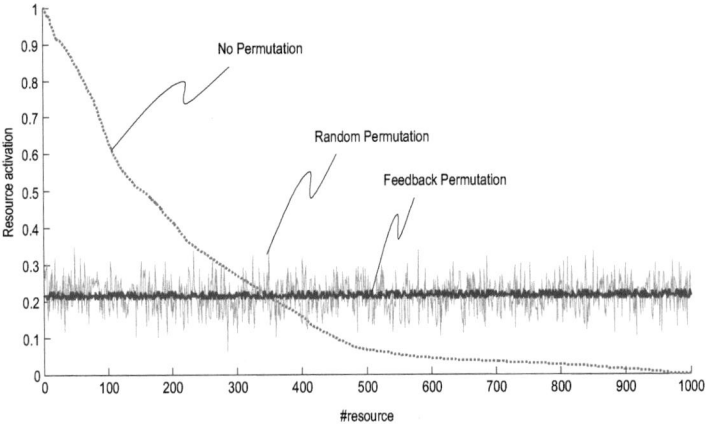

Figure 5.15: Comparison of resource permutation algorithms

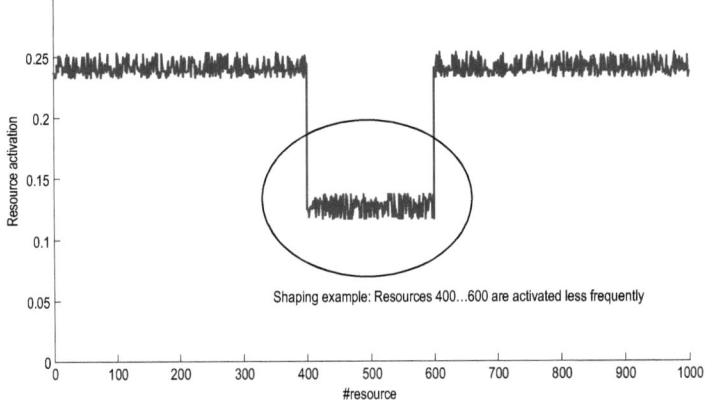

Figure 5.16: The Feedback algorithm allows for shaping the activation probability

The advantage of Random Permutation is that it does not require the activation history, which would need to be stored in a large histogram with N entries. However, it requires the generation of the random resource sequence and a higher permutation rate. In contrast to that, Feedback Permutation depends on the resource activation history, which is memory-intensive. Further, sorting of the resources by the activation field is required, which is computational expensive in case of a large resource set (with high N).

Nevertheless, Feedback Permutation has not only better performance but it also offers the possibility to shape the activation distribution by masking the history data. Shaping of the activation distribution means that the workload is actually not evenly distributed but for individual resources the activation workload can be pre-determined. Figure 5.16 shows an example for shaping. Here, the resources 400...600 are activated less frequently than all others. Since resource operators are compensated for providing control energy to the power grid, some might want to be more active and also gain more compensation as others. This can be realised by masking the history data before determining the new calculation order for the activation levels. Resources that shall be active more often get positive weighting factors < 1, resources that shall be activated less often get positive weighting factors > 1. The past activation values are then multiplied by these weighting factors before sorting. In the example in Figure 5.16, the resources 400...600 are weighted by 2 while all other resources are weighted by 1.

5.3.2 Redundant server system

For the operation of the proposed IRON-CPP system, two general design options are possible: the system is coordinated from a central instance (client-server architecture) [Tan06], or the resources organise themselves in a peer-to-peer fashion [Stein05]. Both options have advantages and disadvantages for a power system related application [Vyv03]. The centralised operation scheme can be advantageous since all operational data is gathered at a central point and can subsequently be used, e.g. for billing purposes. On the other hand, a central server is a single point of failure and also can be a bottleneck for the whole system in terms of connection bandwidth. Therefore, a system with redundant servers should be considered. The peer-to-peer solution scales better, but for this option the activation information of the individual resources is distributed in the system and is therefore not easy to access.

Because of the advantage of central data storage for billing purposes, the option of a centralised approach with a small number ($m > 0$) of redundant servers shall be considered here. These will be called coordinator nodes. The coordinator nodes fulfil two tasks:

(1) They keep track of the activation level arrangement. The activation level arrangement is initially chosen randomly and thereafter modified by a suitable scheduling algorithm (see discussion above). The used scheduling algorithm may rely on the historical data of activation levels. When changed, activation levels have to be communicated to the resource nodes.

(2) They also keep record of the activation of resources in the system. For this, no communication to the resources is necessary. Only the history of S (which is equal for all resources) and the activation level arrangement has to be known for this purpose.

For reasons of failure tolerance, it is necessary to use more than one coordinator node in the system. However, within this ensemble of m coordinators it has to be made sure that only one of these is actually determining a new activation level distribution and sends this to the resource nodes (active coordinator), while the others are acting as passive coordinators and just keep their data structures updated. Otherwise, the coordinators will act counterproductively. Since an active coordinator can go offline (fail) at any time, a solution is needed that hands over the role of the active coordinator to one of the remaining $m-1$ coordinator nodes.

Given the coordinator nodes are uniquely numbered from 0 to $m-1$ and each coordinator node knows all other coordinator nodes, the Bully-Algorithm [Tan06, p. 262] can be used to solve this problem (see Figure 5.17). The coordinator nodes check each others responsiveness regularly. As soon as any coordinator node P detects that another coordinator node does not response, it holds an *election*, which works like this:

(1) It sends an election message to all other coordinators with higher numbers.

(2) If no node answers, P becomes the new coordinator node. It tells this all other nodes. Otherwise, if it receives an answer, it is out of the game.

(3) If a node receives an election message, it sends and OK message back to the sender, and holds another election itself (starting from step 1).

Only one coordinator node will be found in the end which does not receive any answer. This will be the working coordinator with the highest number in the system.

A number of different and more complicated election mechanisms exist, e.g. the Ring-Algorithm [Tan06, p. 263]. However, the actual algorithm chosen for this purpose does not matter, as long as it serves the purpose of selecting a new active coordinator node. This is why just a single example has been given. It can be assumed that the server computers that are used as coordinators for IRON-CPP are equipped with enough computing resources that the complexity of this algorithm does not matter.

Apart from selecting the active coordinator node, a second issue in failure-tolerant duplicated central servers is consistency of the data structures in the servers. As mentioned before, the two central data structures are the activation level distribution (including its history) and the actual resource activation history. However, the resource activation history can be calculated from the record of the $S(t)$ function and the activation level history (compare Figure 5.13). $S(t)$ can locally be determined from frequency measurements and therefore has not to be exchanged between the coordinator nodes. Only the activation level history, which is determined by the scheduling algorithm, has to be kept consistent among all coordinator nodes. Since the activation level distribution has anyway to be sent out by the active coordinator node to all resource nodes in the system once changed, it is the most straight-forward solution to include also the passive coordinator nodes in this broadcast as depicted in Figure 5.18. Thus, the problem of consistency among the coordinator nodes is solved. (Coordinators that went offline for some time will have to contact the other coordinators afterwards to fill their gaps in data history.)

The activation data, which is not measured but calculated on the basis of the known $S(t)$ function, has also be checked against the actual activation measured by the individual resources. Resources can go offline, or the DSM operation can be halted by the resource owner. This additional task has also to be executed on the active coordinator node. In a non time-critical background task, it has to send activation history requests to the resources. The inactive coordinator nodes can read the resource answers as well, and maintain a copy of the activation database of the active coordinator node.

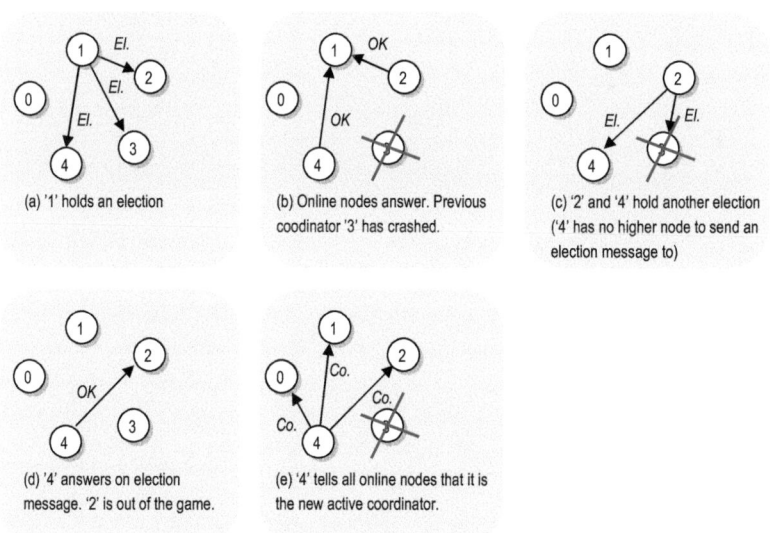

Figure 5.17: Example for the Bully-Algorithm [Tan06, p. 262]. Node '1' detects an offline coordinator. It holds an election. The receivers of the election message hold elections themselves. That node, which does not receive any answer to its election message, becomes the new active coordinator node

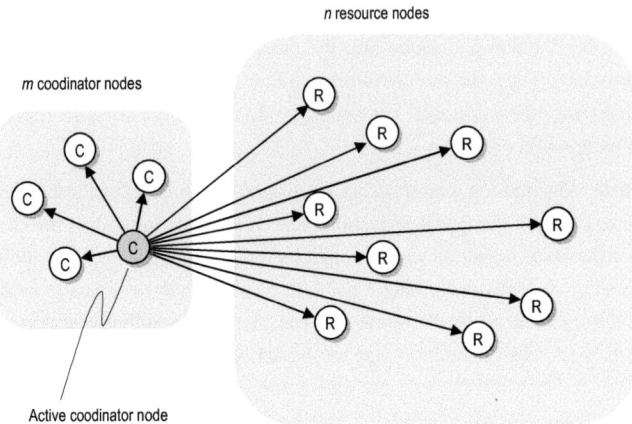

Figure 5.18: Consistency among the m coordinator nodes is achieved by enclosing the passive coordinator nodes into the activation level broadcast, which is primarily targeted to the resource nodes

6. Simulation results and comparisons

For performance evaluation, the IRON-CPP algorithm for primary control energy provision as outlined in the previous chapter is simulated in the time domain with a realistic parameter setup. In a fist step, the primary control power provision is simulated open-loop, meaning that the influence of the system on the grid frequency is not considered. This approach reveals performance and accuracy data for the proposed IRON-CPP algorithm. In a second step, also the frequency feedback over the power grid is simulated and the improvement of frequency control in the power grid using the proposed algorithm is studied.

6.1 Statistical analysis of performance simulations

In the simulations discussed for the statistical analysis, the system reaction on a given frequency series $f(t)$ is simulated without feedback from the DSM system on the grid frequency. (For the dynamic analysis in Section 6.2, this feedback is considered.)

6.1.1 Simulation setup and trajectory visualisation

In simulations, the resource states are calculated according to the equivalent electrical model for a thermal process outlined in Section 4.3.1. The basis of this setup is the parameter set measured from a household refrigerator (Section 4.3.3). Parameters are randomised in order to achieve a population of N DSM resources with related properties. In accordance with the discussion in Section 4.2, the size of the resource population is set to $N = 1000$. This specific resource set is used in all following simulations if not otherwise specified for comparability. A comprehensive description of the resource set, subsequently be referred to as "Resource Set A", can be found in Appendix A.

Activation levels are calculated at the beginning of the simulation and do not change during the simulation run. Activation update has an effect on the utilisation of individual resources but not on the total power consumed by the system, which is of interest here. The amount of control power provided per mHz of frequency deviation (determined by the scaling of the $K(f)$ function) must be chosen in such a way that the maximum value of S can still be covered by the resource set. If the capacity is insufficient due to too less or too weak resources, then clipping effects occur, which result in unintended power consumption behaviour. An example for this is shown in Figure 6.1. This factually means that the system is not providing the control power required in such a case.

The frequency data used in the simulation is taken from a measurement of the actual UCTE grid frequency over several days. The resulting $S(t)$ function is depicted in Figure 5.14 (p. 91) for the first 24 hours of the measurement period.

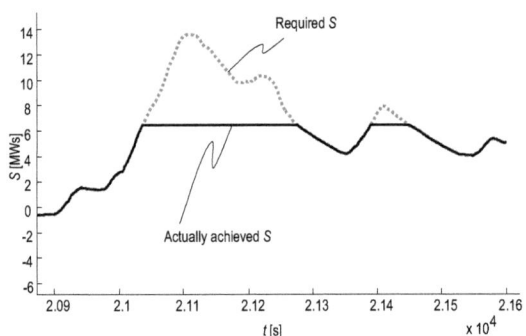

Figure 6.1: Insufficient resource capacity can result in clipping

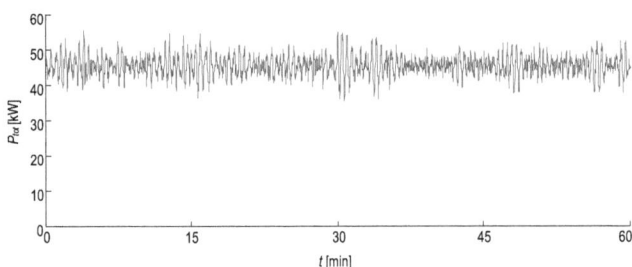

Figure 6.2: Total power over time for Resource Set A and no active control

The simulated total power of the DSM system is shown in Figure 6.2, exemplarily for the first of 72 simulated hours. Switching activity of the thermostatically controlled loads results in a strong noise. Therefore, a statistical approach is chosen for analysis of the simulation results. This is motivated by the trajectory visualisation of the simulation result, which has already been used in Figure 5.7 (p. 83). Now, the trajectory can be drawn with data achieved from actual simulations. The trajectory shows all points in the two-dimensional state space of the system, which have been reached by the system during simulation. The two dimensions are power (total power of all resources) and grid frequency.

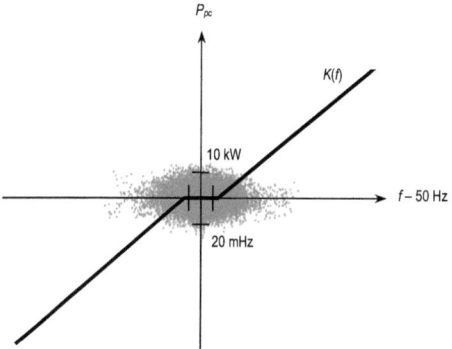

Figure 6.3 System trajectory without active resource control

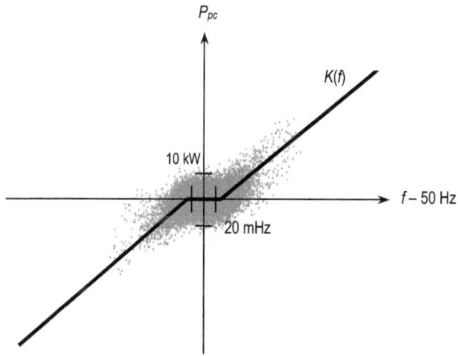

Figure 6.4: System trajectory with active resource control.

The two trajectories shown in Figure 6.3 and Figure 6.4 are obtained from a 72-h-simulation with time step $\Delta t = 15$ s and Resource Set A (see Appendix A). From the power values, the mean value is subtracted and only the deviations to the mean are shown. The first trajectory depicts the system behaviour without any control of temperature set-points. It therefore reflects normal operation of thermostatically controlled loads. No correlation between frequency and power can be observed. The trajectory can also be interpreted as the depiction of the probability density of power and frequency values. The deviation in the power-dimension is caused by the noise from thermostat controls, while the deviation in frequency-dimension is caused by the frequency function used in the

Simulation results and comparisons

simulation. For an ideal control power provider, all values would lie on the $K(f)$ function, which is also shown in the figure.

The second trajectory (Figure 6.4) depicts the case of active DSM resource control. Here, due to the design of the algorithm, power and frequency are correlated. Power values lie within a certain range around the $K(f)$ function. Compared to the ideal result, which is given by the $K(f)$ function, still a considerable deviation can be observed. This deviation however is mainly caused by the noise from the two-state thermostats and not by inaccuracies of the algorithm.

Consequently, the question arises how the performance of the simulated algorithm can be assessed. A technique for quantitative rather than qualitative assessment of the simulation result shown in Figure 6.4 is required.

6.1.2 Quality measure for performance measurement

For a quantitative assessment of the trajectories obtained by simulations, the variance of the simulated total power consumed by the DSM system shall be considered. In case of a system without active control, i.e. no frequency response, the variance is solely determined by the on/off-switching characteristic of thermostat controllers.

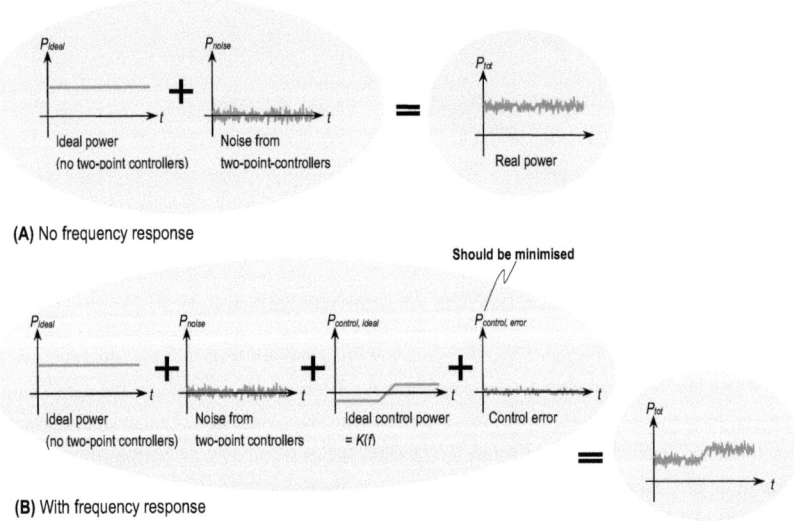

Figure 6.5: Splitting up the sum power signal without frequency response (A) and with frequency response (B) into different Gaussian distributed signals.

Mathematically, each thermostat state i ($i = 1...N$) can be described with a random variable that has the two possible values 0 and $P_{el\,on\,i}$. All random variables are independent, as long no thermal coupling between the individual resources exists (which is generally the case). According to the central limit theorem, the sum of a large number of independent and identically distributed random variables will be approximately normally distributed. For an infinite number it would be exactly normally distributed.) Since the duty-cycles of all resources lie in the same region, it assumed here that the distributions are actually identical. Consequently, the distribution of the total power signal that is the sum power of all N resources is approximately normally distributed. The power consumption $P_{tot,0}$ can therefore be described as a noise-free constant power (sum of all mean power consumptions) plus Gaussian noise (with zero mean) which is caused by the switching operation of the thermostats (see Figure 6.5 A). The index 0 denotes the case of no frequency response:

$$P_{tot\,0} = P_{ideal} + P_{noise} \quad \text{(no frequency control).} \tag{71}$$

In case of a system *with* active control, i.e. frequency response, the variance of the sum power is determined by more than two components, as depicted in Figure 6.5 B. These are:

- The ideal average resource power P_{ideal}
- The noise added by switching thermostats P_{noise} (approximate Gaussian distribution)
- The ideal control power $P_{control\,ideal}$ according to the function $K(f)$ (approximate Gaussian distribution, see Section 5.2)
- The control error power $P_{control\,error}$ caused by non-ideal implementation of the $K(f)$ requirement

All these components add up to the total power with frequency control $P_{tot,1}$ (the index 1 indicates the case of frequency response):

$$P_{tot\,1} = P_{ideal} + P_{noise} + P_{control\,ideal} + P_{control\,error} \quad \text{(with frequency control).} \tag{72}$$

While most components are determined by the simulated environment (resource characteristics, control power demand), the control error power $P_{contro,\,error}$ is a property of the control algorithm and is the variable of interest. The variance of the control error power can be used as a quantitative means of performance assessment. An algorithm that induces less error into the system, resulting in a smaller variance of the total power, is performing better than an algorithm inducing more error and resulting in a higher variance of the total power. It is therefore of interest to calculate the variance of the control error power. A straight-forward approach for calculating the control error power itself is shown in (73):

$$P_{control\,error} = \underbrace{P_{tot\,1}}_{measureable} - \underbrace{P_{ideal}}_{measurable\,(mean)} - \underbrace{P_{noise}}_{unknown} - \underbrace{P_{control\,ideal}}_{known\,from\,K(f)} \tag{73}$$

Simulation results and comparisons

Unfortunately, the noise power cannot be determined from simulation results nor from mathematical considerations since it is not determinable which elements of the thermostat switching are caused by the normal thermostat operation and which elements are caused by the frequency response.

To avoid this problem, not the actual control error power but only the *variance* of the control error power could be calculated. The variance σ^2 is defined using the expectation $E\{x\}$ of a random variable x:

$$\sigma^2(x) = E\{x^2\} - E^2\{x\} \tag{74}$$

However, while *amplitudes* of the individual components listed in (72) can be added to gain the sum power, *variances* of these components can not be added to gain the total variance. The reason for this is that not all processes are statistically independent (75). The control error power is correlated with the noise power generated by the thermostat switches. So, also the determination of the *variance* of the control error power is not possible from a single experiment.

$$\sigma^2(P_{tot,1}) \neq \sigma^2(P_{ideal}) + \sigma^2(P_{noise}) + \sigma^2(P_{control\,ideal}) + \sigma^2(P_{control\,error}) \text{ due to statistical dependence.} \tag{75}$$

Since the control error power neither its variance can be calculated directly from data of one simulation experiment, a quality scale Q^2 is defined referring to *two* experiments, one with and one without frequency response. Apart from active control, both experiments must be made under exactly the same conditions, so that P_{noise} and P_{ideal} are the same (only possible in simulations). The first experiment without frequency response reveals $P_{tot,0}$, the second with frequency response $P_{tot,1}$.

$$Q^2 = \sigma^2(P_{tot,1} - P_{control\,ideal}) - \sigma^2(P_{tot\,0}) \quad \text{with } P_{control\,ideal} = K(f). \tag{76}$$

The quality metric Q^2 compares the power variance of the non-controlled resource set with a corrected power variance of the controlled resource set. The correction removes the influence of ideal frequency response from $P_{tot\,1}$, leaving it with the noise from thermostat controllers P_{noise} and the control power error $P_{control\,error}$.

With the quality metric Q^2, the influence of different parameters of the DSM algorithm can be studied quantitatively. If Q^2 is grater than 0, the frequency response algorithm causes additional variance in the sum power, which should be minimised by the design of the algorithm. If Q^2 is smaller than 0, the control algorithm even reduces the thermostat noise compared to the free-running case.

When Q^2 is determined from simulation results, it has to be taken care that the simulation time step is chosen short enough (1 s or below). A coarse grain simulation time step results in an unintended synchronisation of thermostat state changes, which means that the individual resources are no longer statistically independent. The assumption of statistical independence of the resources can only be made if

$$\Delta t \ll \tau_i \forall i = 1...N, \qquad (77)$$

where Δt is the simulation time step, $\tau_i = b_i^{-1} = R_{L,i}C_i$ is the time constant of resource i and N is the number of resources.

Values of Q^2 are only comparable if they have been achieved with the same resource set and the same time step setting.

6.1.3 Optimisation of τ

The quality metric Q^2 can now be used to determine the optimal choice of the parameter τ of the DSM algorithm. The parameter τ is the time constant of the non-ideal integrator that is used to calculate $S(t)$. The algorithm performance is theoretically ideal if all resources have the same time constant $\tau_i = b_i^{-1}$ and $\tau = \tau_i$ (63). However, in practice the resources have different τ_i. Consequently, the algorithm is not working at its ideal point, and an optimal choice of τ has to be taken (see discussion p. 85).

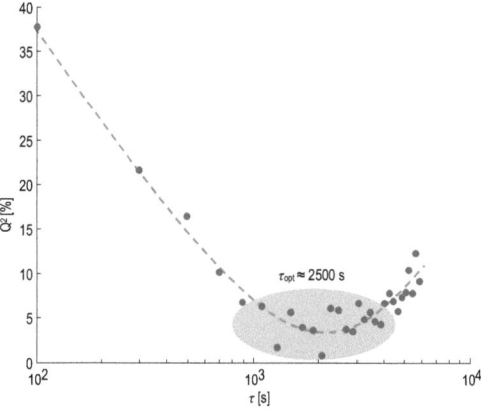

Figure 6.6: Q^2-values for different choices of τ simulated against Resource Set A.
(Q^2 is shown in % relatively to $\sigma^2(P_{tot,1} - P_{control\ ideal})$)

A good rule of thump would be to choose τ equal to the average of all time constants τ_i. The following collection of simulation results confirms this. In these simulations, the algorithm is tested with different settings of τ against Resource Set A (see Appendix A). The Q^2 values achieved are depicted in Figure 6.6. The low-point of the curve is close to the mean of τ_i in Resource Set A, which is 2524 s.

It can also be seen from Figure 6.6, that within a relatively wide band of $\tau = 800...4000$ s the additional variance induced by the non-ideally operating control algorithm is below 8 % of the variance caused by thermostat state changes, which is an acceptable low level.

Not only the mean of the time constants in the resource set, but also the distribution of $\tau_i = b_i^{-1}$ within the available set of resources has an influence on the performance of the algorithm. Simulations with different resource sets reveal that the more similar the individual resources are, the better the algorithm performs. The results shown in Table 3 are for a simulation with $N = 1000$ resources and time step $\Delta t = 1$ s. The resource sets used are described in detail in Appendix A.

Table 3: Power variances of different resource sets

Resource Set	Distribution of τ_i [s]	$\sigma^2(P_{tot\,0})$ [MW²]	$\sigma^2(P_{tot\,1} - P_{control\,ideal})$ [MW²]	Q^2 [MW²]
C	2000...3000	4.6903	4.7886	0.1883
A	1500...3500	4.8004	5.2033	0.4029
B	500...5500	4.7146	5.4460	0.7314

The reason for the better performance is that the algorithm is actually designed for resources with similar time constants $\tau_i = b_i^{-1}$. The error term in (59) is only disappearing if all b_i are equal. The closer the resource set is to this condition, the better the algorithm performs. However, a certain range in the deviation of time constants is acceptable. If in a real DSM system resources with very different time constants have to be integrated, they should be grouped by similar time constants and for each group of resources a separate instance of the DMS algorithm with separate activation levels and τ setting should be used.

6.1.4 Slow resource reaction

IRON-CPP utilises a dedicated scheme for coordinating the activity of individual resources in order to achieve a smooth control power unction. As outlined earlier, each individual resource has a discrete operational state: positively charged, neutral and negatively charged. Values between these discrete states are only reached during state changes. A state change triggers a change in resource power consumption for some time. Ideally, after this time the next resource should be activated immediately, so that the next resource releases the previous seamlessly. But this is not always the case.

As shown in Figure 6.7 A, resources state-changes are caused by the S value crossing the activation level of the respective resource. Due to the quantisation of the S range into discrete activation levels, and also the quantisation of resource states, the resource reaction after the activation is not depending on the slope of the S function. Consequently, three different cases can occur:

- The resource reaction is just right and it is finished *exactly* when the next resource is activated (Figure 6.7 B).
- The resource reaction is too short (but the amplitude too high) and it is finished *before* the next resource is activated (Figure 6.7 C).
- The resource reaction is too long (but the amplitude too low) and it is finished only *after* the next resource is activated (Figure 6.7 D).

In order to determine the impact of this effect on the algorithm performance, a series of simulations is conducted between which the capacity value of the resources is varied. The basis for this simulation is Resource Set A (see Appendix A), but the C value of each resource is changed so that the range of τ_i is achieved as shown in Figure 6.8.

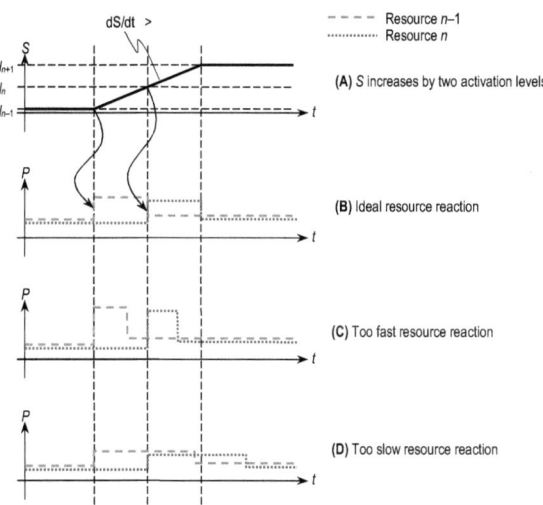

Figure 6.7: Seamless and non-seamless resource reaction

The resulting curve can be partitioned into four different sections (Figure 6.8).

(A) For very small capacities, the resources react too fast, which causes significant distortions. This problem could be solved by alterations of the local resource controller. It would have to detect the slow change of S over time and also change the set-point temperature of the resource slower in accordance to S.

(B) For an optimal τ_i, the Q^2-value is minimal. Resource reactions are ideal.

105

Simulation results and comparisons

(C) For a further increase of the value of C (and τ_i respectively), the curve is flat, but increases slightly to the right. Here, resources react already too slow, but reactions overlap each other so that no significant gaps occur.

(D) For very high values of C, resources become too slow. They finally show nearly no reaction at all. Here, the performance degrades.

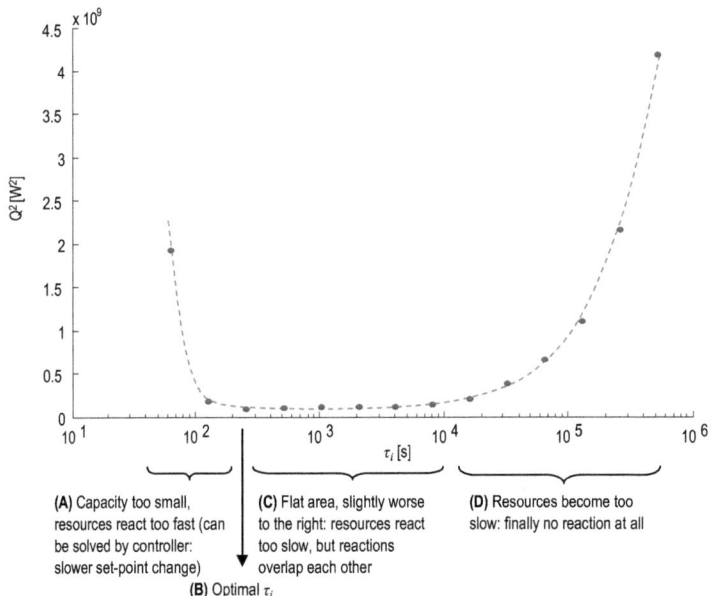

Figure 6.8: **Performance effect of different resource reaction times**

The algorithm should be operated only in the flat area (C), including the optimal point (B). This can be achieved by avoiding exceptionally fast and exceptionally slow resources in the resource set.

6.1.5 Influence of storage discharges

The quality measure Q^2 can also be used for studying the influence of storage discharges. Storage discharges occur whenever the energy stored in the resource (which acts as a conceptual energy storage) is drained by an external event or demand. If the resource is a cold storage room, and new goods are transported into the room, which are warmer than the inside temperature, then this is a storage discharge. If the resource is a water boiler, and the hot water is used for a warm bath, then this is also a storage discharge.

106

Storage discharges occur more or less frequently and with more or less impact. The impact refers to how much energy is detracted from the resource. Frequency of occurrence and impact strongly rely on the type of resource and its usage. In order to cover the large range of different real-world situations, in this work a simplified model of storage discharge is implemented. In the simplified model, the storage discharge can happen during any time step of the simulation with a certain probability $p_{discharge}$, which is specified in average number of discharges per day. If a discharge happens, it has always maximum impact, i.e. the resource is completely drained. This is realised by resetting the temperature state of the resource, which reflects the difference between inside and ambient temperature, to zero. Figure 6.9 depicts an exemplary temperature plot over time for one resource with 8 discharge events in 4 hours.

Even though a time-dependent discharge probability would be closer to reality, here a constant probability is used because for statistical analysis over several days the actual time of a discharge events does not play a role, as long as not all discharges happen at the same time. Equally distributed and maximum-impact discharges are a worst-case model for resource drains caused by external events or demands in regard to their impact on the power deviation. This is because in the equally distributed case, the individual discharge events to not overlap each other. So, each single event does fully influence the power deviation; no shadowing effects occur.

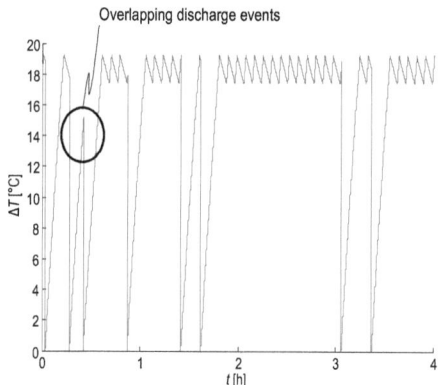

Figure 6.9: Resource temperature over time with random discharge events

In a simulation experiment the impact of discharges on the performance of the DSM algorithm is evaluated. In a series of simulations with different discharge probabilities, the variances required for calculating Q^2 are recorded (see Figure 6.10). For each point, 10 simulation runs are performed and the mean of the variance results is shown. This averaging of results is necessary due to the fact that each simulation is now a probability experiment, and without a number of repetitions, results would not be comparable. Simulations are performed against Resource Set A with a time step of 2 s.

Simulation results and comparisons

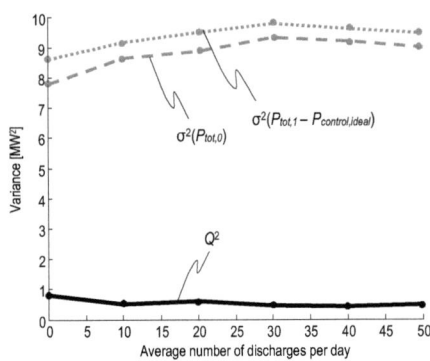

Figure 6.10: **Additional power variance due to discharge events**

Both variances, $\sigma^2(P_{tot,0})$ and $\sigma^2(P_{tot,1} - P_{control,ideal})$, are first increasing with the average number of discharges per day and then declining, after reaching a weak maximum at about 30 discharges per day. The increase is due to additional switching activity caused by the discharge events. However, when a critical density of discharges per hour is reached, the re-charge curves begin to overlap each other (for am example see second and third event in Figure 6.9). This results in a reduction of switching activities and therefore also in a reduction of variance. The difference of the two variances, which represents the error of the DSM algorithm, stays approximately constant.

From this experiment it can be concluded that the effect of resource discharges has no significant effect on the DSM algorithm performance. Rather than causing additional variance, Q^2 is even slightly improved by discharges in comparison to the case of no discharges.

While discharges are not affecting the DSM algorithm, they certainly have a significant effect on the total power consumed by the resources in the system. However, this fluctuation of the resource power is not of interest for the control power provision. The changing discharge probability contributes to the fluctuation of the power demand over the day, which it is a known effect and is handled by day-ahead power plant schedules. Since the discharges have no or only minor effects on the power deviation, this effect can be counted to the basic resource power, which is modelled by the parameter c in (43) (p. 72).

6.2 Dynamic analysis

In the previous section, the open-loop characteristics of the IRON-CPP algorithm have been evaluated by a variance-based metric. In a second step, the complete control circuit spanning from power grid frequency control to the DSM system is examined. The power system reaction on a generator failure is simulated and system reactions with conventional and DSM control power provision are compared.

6.2.1 Simulation scenarios and underlying models

For the dynamic simulation, a national power grid is modelled. An overview of the simulated components is given in Figure 6.11. The underlying models used are described in Chapter 2.

Generators and loads in the power grid are simulated by aggregate generator and load models. On the generation side, the generator G1 fails at $t = 0$. The system reaction on this failure is simulated for three different scenarios: "AT", "UK" and "UK combined". Austria has been selected because here the IRON-CPP algorithm was developed. The UK was chosen because of the comprehensive modelling data available for the UK power grid from [Short07]. The scenarios differ in the size of the simulated power system. In all scenarios, no power exchange with neighbouring control zones (countries) is considered in order to prevent the simulation from becoming too complex. For the UK scenarios, this is realistic since the UK power system is only connected via DC connections to the European continent. However, for the Austrian power grid, this is a strong simplification since the Austrian power grid (APG) is part of the UCTE and therefore embedded into European power exchange.

For simulating the DSM resources, a population of $N = 1000$ refrigeration systems according to the distributed resource model discussed in Section 4.5 are simulated. The total power of this resource set is amplified in order to represent a larger set without increasing the simulation complexity. The amplification factor depends on the average DSM power required in the given scenario.

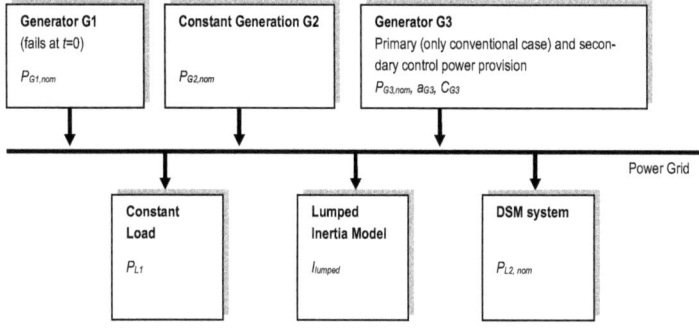

Figure 6.11: Components of the dynamic analysis. Model parameters are given in

Table 4.

Table 4: Simulation scenarios for the dynamic analysis

Symbol	Value in scenario			Description
	AT	UK	UK combined	
$P_{G1,nom}$	150 MW	31 GW	31 GW	Power generated by G1 that fails at $t = 0$
$P_{G2,nom}$	6850 MW	1320 MW	1320 MW	Power generated by G2 (lumped model for remaining generation in grid)
$P_{G3,nom}$	1000 MW	5 GW	5 GW	Power generated by control power generator G3 @ 50 Hz
a_{G3}	-500 MW/Hz (conventional) 0 (DSM)	-2.5 GW/Hz (conventional) 0 (DSM)	-2.5 GW/Hz (conventional and DSM)	Slope of the droop characteristic of G3. In AT and UK scenario, primary control is deactivated when simulating the DSM load
C_{G3}	6×10^5	3×10^6	3×10^6	Weighting factor for secondary control
I_{lumped}	648.5 kWs	3.025 MWs	3.025 MWs	Inertia constant of the lumped inertia model
P_{L1}	8000 MW (conventional) 7850 MW (DSM)	37.320 GW (conventional) 36 GW (DSM)	37.320 GW (conventional) 36 GW (DSM)	Constant load power.
$P_{L2,nom}$	0 (conventional) 150 MW (DSM)	0 (conventional) 1320 MW (DSM)	0 (conventional) 1320 MW (DSM)	Mean power of the DSM load @ 50 Hz

In the "AT" scenario, the total load is assumed to be 8 GW (maximum load, winter day, [UCTE05]). According to [Bra06, p. 56], the domestic refrigeration load is 180 MW. Basing on this figure, the DSM load (thermal load processes equipped with a DSM interface) is assumed to be 150 MW. On the generation side, the total 8 GW are covered. The main share is committed by G2, which is a constant aggregate generator model. Generator G3 models the aggregate control power provision in the Austrian power grid. The generation power of G1 is chosen to match the DSM load. For the AT scenario, the conventional system behaviour with primary control power from G3 is compared to the DSM system behaviour with primary control power from the DSM system.

In the "UK" scenario, the values for total power, generation and refrigeration load are chosen according to the simulations performed by Short et al. [Short07] in order to achieve comparability. Also in this scenario, the conventional system behaviour with primary control power from G3 is compared to the DSM system behaviour with primary control power from the DSM system.

The "UK combined" scenario differs from the "UK" scenario in only one respect: Here, conventional primary control power provision is not replaced by DSM operation, but conventional and DSM control power provisions are combined. While in the other two scenarios, the ability of the

DSM algorithm to replace conventional primary control power provision completely is examined, the "combined" scenario studies the combination of both.

All parameters of the scenarios are listed in Table 4.

6.2.2 Result of dynamic simulation

The simulation results for the IRON-CPP algorithm in scenario AT are depicted in Figure 6.12 A-C. In Figure 6.12 A, the power behaviour of the conventional system (i.e. no DSM operation) is shown. After G1 fails at $t = 0$, fast activation of spinning generation reserves, which is the primary control, takes over the loss of generation. The primary control action is then replaced by secondary response, which is in full operation after about 10 minutes. 15 minutes after the failure event, a replacement generator is beginning to start up. After 36 minutes, the loss of generation is completely replaced by the replacement generator (tertiary reserve).

Figure 6.12 B, the case of primary control provided by the DSM system instead of G3 is shown (IRON-CPP algorithm). Here, the control power originates from a reduction in load rather than an increase in generation. The DSM resource power (black solid line) decreases after the generator failure to about 50 % due to set-point change in thermal processes. Secondary control compensates the remaining 50 %. The DSM power is then increasing over the next 30 minutes, because the thermal storages are drained and thermostats one by one switch on again. The DSM power even rises above its former average level of 150 MW. This is the rebound effect: the drained storages recharge.

Figure 6.12 C compares the grid frequency for the conventional and the DSM case. The area of interest is the frequency behaviour immediately after the 150 MW generator failure. In the conventional system, the frequency drops because energy is taken from the rotating masses of the remaining generators. Primary response restricts this frequency drop to 49.7 Hz. Secondary control then restores the nominal frequency. In the DSM system, the frequency drop is more severe (49.3 Hz) because the IRON-CPP algorithm does not react as fast as conventional control energy provision does. The reason for that is that the $S(t)$ variable, which determines the DSM system power, is not changing fast enough. It is essentially calculated by a integration of frequency error, and some time is needed until $S(t)$ becomes large enough that a high number of resources switches off. However, once the IRON-CPP system has reacted, the frequency is restored faster than in the conventional case.

The increased frequency instability in the DSM case, which can be observed in Figure 6.12 C, is not only caused by the IRON-CPP system but by the fact that the conventional system is simulated with the thermostatically controlled loads replaced by a constant load (in order to make the results comparable to [Short07]), resulting in a smoother frequency result.

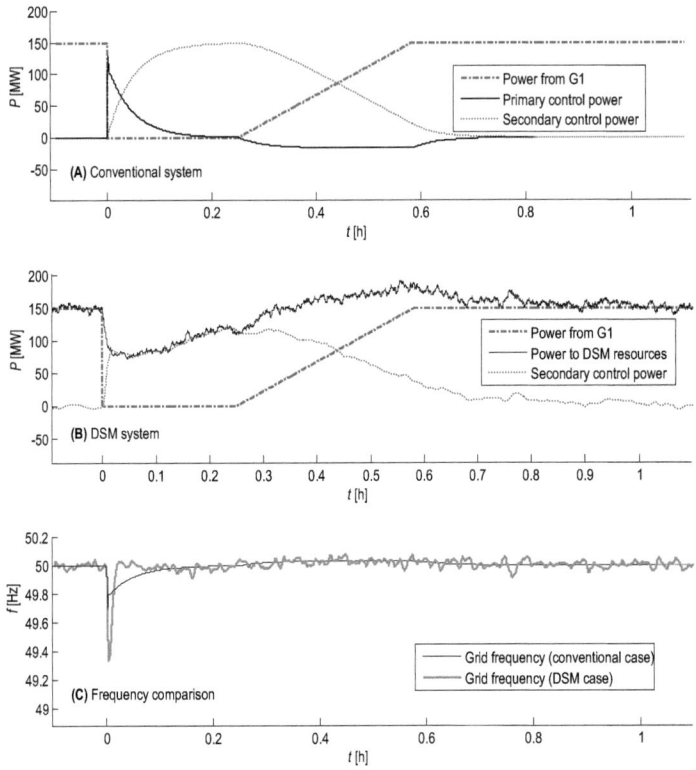

Figure 6.12: Simulation of IRON-CPP, scenario AT

6.2.3 Improvement of the dynamic behaviour

The slow reaction of the IRON-CPP algorithm on a fast frequency drop as it can be observed in Figure 6.12 C is undesirable. It results in severe frequency drops and an earlier utilisation of secondary reserves compared to the conventional system. Secondary reserve is also needed for restricting the frequency drop in this case, for which it is actually not intended [UCTE04a].

As already mentioned, the slow reaction of the IRON-CPP algorithm is due to the integral nature of the calculation of $S(t)$. It can be seen from (78), that a drop in frequency does not have an immediate impact on S. Only after a number of time steps, S changes considerably.

$$S(t+\Delta t)=\underbrace{\Delta t K(f(t))}_{\text{frequency error}}+\underbrace{S(t)\left(1-\frac{\Delta t}{\tau}\right)}_{\text{integrator with losses}} \tag{78}$$

For improving the reaction speed, a proportional part with weighting factor k_{pi} is added to (78):

$$S_{PI}(t)=S(t)+\Delta t k_{PI} K(f(t)) \tag{79}$$

The new time function $S_{PI}(t)$ replaces $S(t)$ in the algorithm. However, it is based on $S(t)$ since the proportional part is not integrated over time. Such improved algorithm is called "IRON-CPP-PI". Figure 6.13 shows the calculation method of $S_{PI}(t)$ in a block diagram and also the simulated $S(t)$ and $S_{PI}(t)$ for the scenario AT. Due to the new proportional part, $S_{PI}(t)$ falls from about 0 to −3 MWs immediately after the generator G1 fails at $t = 0$.

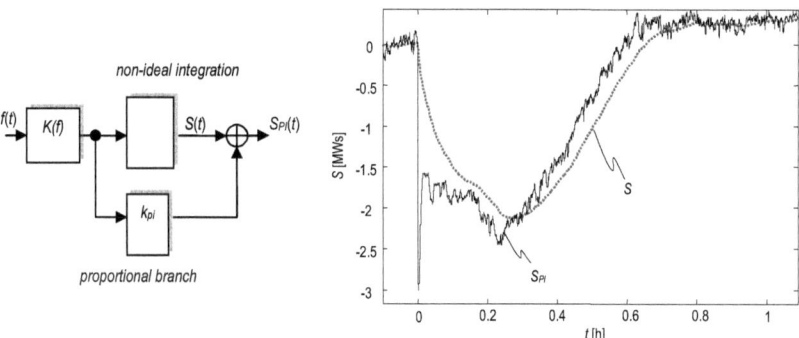

Figure 6.13: Comparison of integral and proportional/integral calculated S

The simulation of the new IRON-CPP-PI algorithm reveals better results. As shown in Figure 6.14, about 80 % of the DSM load is freed very soon after the generation drop (compared to 50 % for IRON-CPP). IRON-CPP-PI keeps the frequency within 0.2 Hz deviation from 50 Hz and performs new better than the conventional system.

Compared to the conventional case, IRON-CPP-PI also relaxes the timing requirements for secondary control. The activation of the full secondary reserve is delayed by the DSM system, which provides control power faster and longer than conventional primary control. The faster response time can be achieved because instead of a few large and inert generators, many small and flexible loads provide the control power in this case.

Simulation results and comparisons

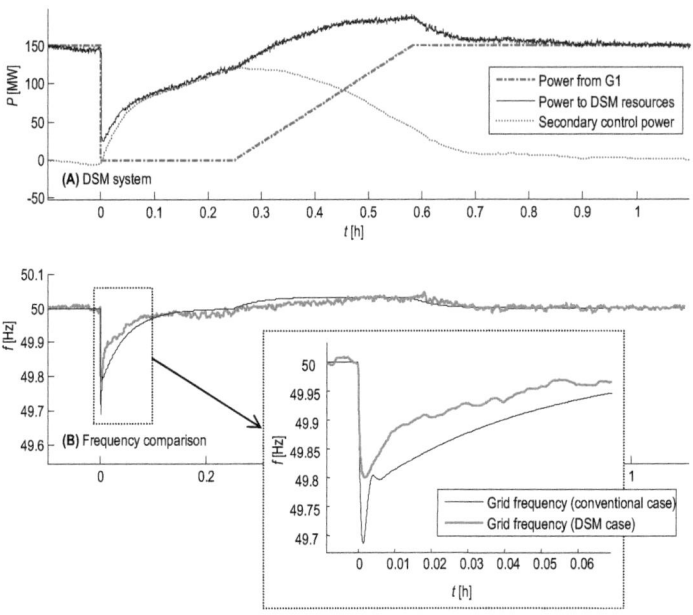

Figure 6.14: Simulation of IRON-CPP-PI, scenario AT

6.2.4 Comparison with the approach of Short et al.

The DSM control algorithm proposed by Short et al. [Short07] has already been discussed in Section 3.3.2. It is an alternative to the proposed IRON-CPP-PI algorithm. It works without communication between the DSM resources, which is cost-effective but does not allow for activity-based refund of primary control power provision. Refrigeration systems, especially household refrigerators are targeted. The thermostats are controlled according to the grid frequency. The individual temperature set-points are changed proportionally to the frequency deviation.

$$T_{set,DSM} = T_{set,norm} - u(f - f_{nom}) \qquad (80)$$

The factor u is chosen 5 °C/Hz according to [Short07] (compare Figure 3.5 p. 39).

Simulation results of the algorithm of Short et al. are shown in Figure 6.15 for the AT scenario. The algorithm performs well in terms of fast reaction on the frequency drop. Since all thermostat set-points are immediately changed, the DSM load reduces to about 20 % of its steady-state level at 50 Hz. This results in fast frequency regeneration. The frequency rises for a few minutes. Then however, refrigerators begin to switch on again. Soon, the frequency is stable and then even falls again due to the rebound effect. This second frequency fall is then stopped by secondary control. A further increase of secondary control power and the start of the replacement generator result in a considerable overshoot, so that the frequency rises to 50.1 Hz before it reaches the nominal value again.

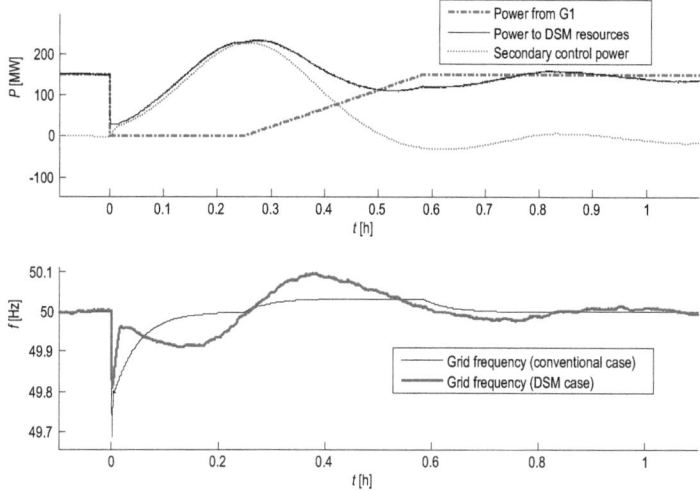

Figure 6.15: Simulation of algorithm after Short et al., scenario AT

The frequency behaviour of this approach is undesirable[1]. Only in combination with a supporting primary controller, the Short et al. algorithm performs acceptable. The main issue is the linear set-point shift, which does not take into account the DSM storage behaviour. As discussed in Section 4.3, the resource power is primarily depending on the temperature set-point *derivation* and not its absolute value. The proportional extension of the IRON-CPP algorithm has shown that proportional elements in the frequency-power relation improve the short-term system answer on frequency drops, but they do not have a stabilising long-term effect as integral elements (which match the have derivation on the resource side) have. The algorithm suffers from this design fault.

6.2.5 Comparison with the GridFriendly Appliance Controller

Another alternative to IRON-CPP-PI is the algorithmic approach of the GridFriendly Appliance (GFA) Controller [Ham07], which has already been discussed in Section 3.3.1. The algorithm also does not make use of any communication among the DSM nodes. The approach here is a cost-

---

[1] The frequency curve Fig. 8 in the publication of Short et al. [Short07] suggests better frequency behaviour. Despite their well-documented simulation approach, the author was not able to reproduce this result. After careful consideration, the author assumes an error in Fig. 8 of [Short07].

effective single-chip solution that can be integrated into any household appliance and improves grid stability.

This algorithm is threshold-based, i.e. it does only become active if the grid frequency has crossed a certain lower frequency threshold. The appliance power is then reduced. In the simulation of the GFA system presented here, the appliance is a refrigerator and load reduction is achieved by temperature set-point change. In order to avoid all appliances to react at the same time, the algorithm depends on randomised reaction delays [Ham07, p. vii]. Once the frequency enters a certain hysteresis area, which lies below the nominal frequency, a reaction delay timer is started. This timer activates the load reduction. Once the frequency leaves the hysteresis area, another timer is started, which ends the load reduction after a randomised time constant. Frequency threshold and reaction times are individual in each resource. The time constants and randomisations used in simulations here are taken from [Hamo07, p. 1.6–p.1.7] and are listed in Table 5.

Table 5: Parameters of the GFA simulation

Parameter	Nominal value	Variation
frequency threshold	49.9 Hz ... 49.95 Hz	equally distributed
frequency hysteresis	0.02 Hz	none
reduction delay	0.4 s	none
re-activation delay	16 s ... 32 s	equally distributed
Set-point change for load reduction	1 °C	none

The simulation result of the GFA approach is depicted in Figure 6.16. In absence of a conventional primary control provider, the algorithm is able to restrict the frequency to values larger than the lower frequency threshold, apart from a few seconds after the generator failure, where all reaction timers are still running. The DSM load is nearly reduced to zero at the beginning. The frequency stays close to the threshold level as long as DSM resources are recharging. Only after the recharge phase is finished in all resources, the frequency rises. Here however, the first major drawback of the GFA algorithm becomes eminent: it is not designed for avoiding over-frequencies. Consequently, the grid frequency resumes with numerous overshoots.

Even if the GFA algorithm was not designed to replace conventional primary control power, this simulation result brings the design faults of this algorithm to light. First of all, it follows a very simple "if frequency is too low then switch off" scheme and does not make use of inter-resource communication, which potentially could be used to avoid instability caused by load behaviour synchronisation. The algorithm is aiming to improve grid stability, but the basic approach even worsens stability because it synchronises load reactions. In order to avoid this, instead of changing the basic approach the symptoms are cured by randomisation.

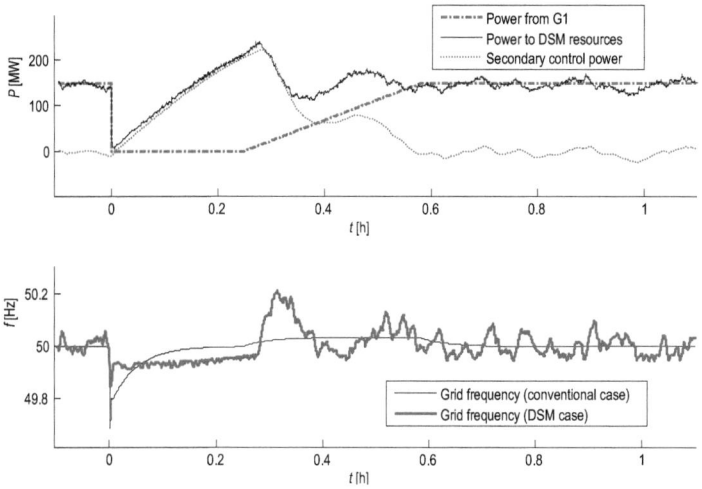

Figure 6.16: Simulation of the GFA algorithm, scenario AT

Secondly, the algorithm is designed without any particular load model kept in mind. It is simply assumed that the resource power can be reduced by a certain amount. This is basically load shedding without energy storage in the demand side process. The advantages of demand side storage (load management without energy service reduction) are neglected by the algorithm design. Rebound effects, which occur when combining the GFA controller with demand side storages, have not been considered and this results in severe oscillations of the grid frequency.

6.2.6 Discussion

The conventional primary control energy provision as well as the algorithm of Short et al, the GFA controller, the IRON-CPP and the IRON-CPP-PI algorithm are simulated for the three scenarios AT, UK and UK combined, which are described in Section 6.2.1. The UK combined scenario has been specially introduced to allow a fair comparison with algorithms that are not specifically designed to replace conventional primary control energy provision (GFA controller).

As a metric for dynamic performance the integral frequency error is defined:

$$e_f = \frac{1}{t_{sim}} \int_0^{t_{sim}} \left(\frac{f(t) - f_{nom}}{f_{nom}} \right)^2 dt \qquad (81)$$

Simulation results and comparisons

The lower this unit-less error value is, the better the algorithm performs. The simulation of the conventional power system acts as reference scenario.

The simulated integral frequency errors of all considered algorithms in all considered scenarios are shown in Figure 6.17. It can be seen that the original IRON-CPP approach is only feasible in combination with conventional primary response (scenario UK combined). In the other scenarios, the frequency drop is too severe due to the slow load reduction after the generator failure.

The approach of Short et al. is better than IRON-CPP in the scenarios AT and UK. The main reason for this is that the frequency-proportional set-point control has a reducing effect on the thermostat noise generated by the resource set, resulting in a smaller integral error than the "noisy" IRON-CPP algorithm without a proportional component.

The GridWise GFA system achieves the highest error values. Even in the combined scenario, the system performs not better than the conventional power system. The reason for that is that the rebound effect caused by GFA-triggered load reductions leads to severe oscillations of the grid frequency.

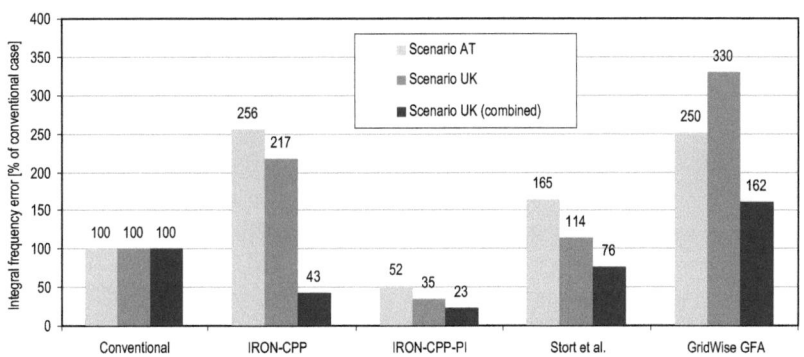

Figure 6.17: Performance comparison of frequency-response algorithms

Best results are achieved by the IRON-CPP-PI algorithm. In all three scenarios, it reaches better frequency control than the conventional solution. The appropriate underlying resource model is a key factor for this result. The algorithm is specifically designed to replace conventional primary response by management of energy in demand side processes. It combines proportional and integral reaction on frequency deviations (PI control) and therefore has a good impulse answer and steady state behaviour. It also relies on communication between the resources, which enables a better coordination of individual resources but is causing additional costs.

7. Implementation

While previous examinations have studied the proposed IRON-CPP algorithm by means of simulations, the implementation of system key komponents in hard- and software is discussed here. This implementation was motivated by several questions: How complex is the infrastructure needed to execute the proposed algorithm? How can the communication between the nodes actually be realsised? What costs for the hardware have to be expected in case of high-volume production? How can actual interfaces between communication node and electrical load look like?

One question, however, cannot be answered by this prototype-level implementation: the question of the acuracy of the previously conducted simulations. This could only be answered if a significant number of electrical loads were equipped with the load management hardware. However, only six units have been realised. Comparisons between the real system and simulation results can only be made on the level of individual resource behaviour, a study that has already been discussed in Chapter 4, Section 4.3.3.

7.1 System design considerations

The overall system, of which parts were implemented, is shown in Figure 7.1. Following the discussion about centralised or peer-to-peer architecture in Section 5.3 (Workload balancing by activation level update), the decision for the central server option has been taken. The upper part of the depicted system (power grid, generators and loads) already exists. The communication network is also considered as available technology; only a selection of adequate communication technology has to be done. Remaining parts for implementation are the DSM interface unit and the central server.

With only a few resource nodes in the field, the central server of the system is not strongly loaded by the task of activation level update. In fact, its main task in this prototype setting is to gather measurement data which is collected by the field nodes and then sent to the central server for further analysis. Information about network availability of the field nodes, temperature set-points and power consumption of connected electrical loads are of interest here. A redundant realisation of the central server does not make much sense in this setting; apart from testing the election algorithm (Section 5.3.2). Therefore, a single-server solution was chosen.

Implementation

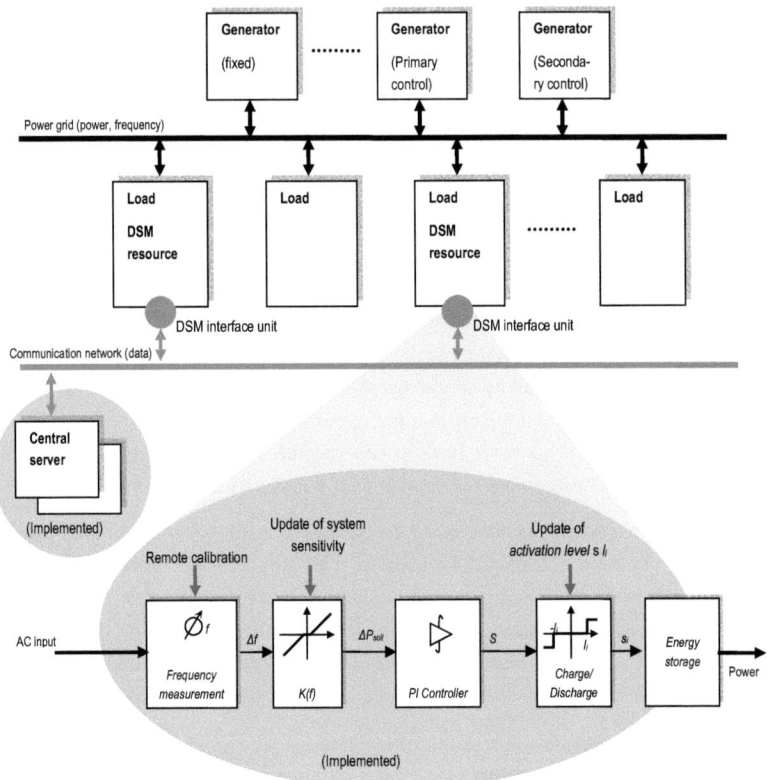

Figure 7.1: System overview. Resources are connected by two networks: power grid (above) and communication network (below). The DSM interface unit links the load with the communication network.

For the communication infrastructure, Internet communication has been chosen for its ubiquitous availability. The central server is set up as a dedicated machine in a server room, having broad band connection to the Internet via the campus network. The field nodes could use different Internet access technologies. Potential options are:

- Direct Ethernet connection
- Modem dial-in
- WLAN (Wireless Local Area Network)
- GPRS (General Packet Radio Service)
- HSDPA (High Speed Downlink Packet Access)
- PLC (Power Line Communication)

and others. Once Internet connectivity is achieved, the communication between resource node and central server can be realised using TCP or UDP Internet protocols. On top of these, a dedicated DSM application protocol has to be defined (see Section 7.3).

In the majority of cases the resource node will be located behind an Internet firewall or even Proxy server, and will be assigned dynamic IP addresses. Thus, it is not possible for the central server to contact the field nodes, since the field node Firewall would block such an external connection attempt. The field nodes rather have to initiate the connection themselves. In the implemented experimental setup, the DSM interface units (field nodes) use TCP/IP connections, which are initiated on power-on and maintained as long as the network is available. Once the connection is broken, the field node tries to reconnect. The server keeps track of all connections, their initiation and brake offs. Thus, the availability of the network connection can be tracked.

For the final system, it is not necessary that all field nodes maintain their connections at all times. However, they will have to connect the central server(s) in regular time intervals for requesting the activation level update. For testing the experimental server under heavy load, i.e. more than a few resource nodes connecting to it, DSM interface units in the field can be simulated by similarly behaving client software.

7.2 DSM interface unit

The DSM interface unit realises the interface between the DSM resource (e.g. load) and the DSM network. The implemented prototype serves as hardware equipment for experiments as well as a basis for estimating the costs for a commercial production. While the central server hardware is a standard off-the-shelf product, and only the server software has to be developed, the implementation of the DSM interface unit requires hardware and software development.

Figure 7.2: Final design of the DSM interface unit

Implementation

The requirements for the DSM interface unit are:

1. **Provision for load management**: an electrical load should be influenced either by interrupting its power supply directly, or by changing operational parameters like temperature setpoints. For both purposes, a relay contact with sufficient current rating must be provided.
2. **Provision for state observation of the connected electrical load**: The unit must provide data inputs that allow for connection of sensors for state observation of the connected electrical load. These may be used for temperature measurements etc.
3. **Provision for frequency measurement**: the grid frequency needs to be measured by the device with sufficient resolution.
4. **Provision for power measurement**: The power consumed by the connected electrical load must be monitored by the DSM interface unit. This feature is needed for analysis of the load management algorithm as well as for calculation of the energy storage capabilities (capacity and isolation resistance) of the connected electrical load.
5. **Provision for Internet connectivity**: The DSM interface unit must be able to connect to the Internet using an appropriate access technology (i.e. one that is available at the location of the interface unit, e.g. WLAN or GPRS).
6. **Provision of computing power**: A programmable microprocessor must be available that can be programmed to execute the algorithm for primary control energy provision as discussed in Chapter 5.

The actual system design was preceded by a dedicated review of existing products to answer the question, if there any ready-made system that can be adapted or re-used. Of course, no answer with 100 % confidence can be given since it is not practically possible to gain a complete and exhaustive overview of all available products in the context of DSM interfaces. However, the results gained from this product review suggest that there are no products that fulfil even the most important parts of the specification, apart from generic and modular industrial PCs, which could be extended by dedicated modules that provide the required services. This however would result in a expensive and bulky solution. Therefore, the decision for a new hardware design for the DSM interface unit was taken.

The actual implementation was conducted in seven steps:

1. **Internet access technology selection**: From the large variety of potential access technologies, three mainstream technologies were selected and compared in a practical test: GPRS, modem dialup and WLAN. The comparison focussed on aspects communication (achievable data throughput, typical latency, duration of connection initiation), costs (hardware costs, connection costs) and ease of installation. The results of the comparison are listed in Table 6. First prototype boards were developed for the comparison, one for each access technology. These proto-

types consist of a microcontroller, the Internet access module (WLAN, GPRS, or modem), relays and power supply. Due to its complicated configuration process during installation and the high connection costs, the telephone modem option has bee abandoned. For the final prototype design, it was decided to make provisions for both a WLAN card and a GPRS modem (alternative options).

2. **Form factor considerations**: Depending on whether the target load is used in an industrial or private home context, different form factors for the DSM interface unit are preferable. Since the proposed system mainly targets large-scale temperature control, pumping, ventilation or heating systems, the interface unit will probably be situated in a switching box. Therefore, a standard DIN-rail casing was selected (see Figure 7.2).

3. **Processor selection**: The proposed algorithm does not require extensive computing resources. For keeping the costs as low as possible, a simple 8 bit microcontroller was chosen (Atmel ATmega 128 @ 8 MHz). For the prototype version of the DSM interface unit, this processor provides more than enough capabilities. However, in the final system cryptographic algorithms have to be used to prevent that unauthorised parties gain access to measurement data or even control over electric loads. (A cryptographic protocol for securing field nodes from malicious attacks that potentially can be used here can be found in [Nae06].) These algorithms might need more computing power. Nevertheless, since the communication delay is not significant in the proposed control algorithms, more resources are mainly needed in terms of memory, not in terms of processor speed.

4. **Selection of remaining components**: The remaining components were selected around the "fix-points" of processor and communication interface. A complete overview of all realised functional blocks of the DSM interface unit can be found in Figure 7.3.

    For *measuring the resource state*, six insulated digital opto-coupler inputs are provided. Changes on these inputs can trigger a processor interrupt. Continuous states have to be transferred to discrete representation (e.g. for temperature values: above or below threshold) by external circuitry. A direct temperature sensor input (PT100 or PT1000) should be added in a future version.

    For *frequency measurement*, another opto-coupler is used to insulate the 230 V sine wave signal from the low voltage circuitry. The opto-coupler produces a rectangular signal (TTL level) with the same frequency as the grid input. The remaining signal processing is done by the microcontroller, resulting in a low cost solution consisting only of the opto-coupler and a few resistors. The oscillation period $T = f^{-1}$ is measured by one of the hardware timers/counters provided by the microcontroller. The counter frequency is 625 kHz, so that a counter result of $n = 12500$ indicates a grid frequency of exactly 50 Hz. The resolution achieved by this approach is 4.00032 mHz in the 50 Hz working point. The measurement error is determined by the crystal tolerance, which can be assumed to be 120 ppm in the worst case [10]. Figure 7.4 shows the achieved accuracy, which is bounded by +/−2 digits.

Implementation

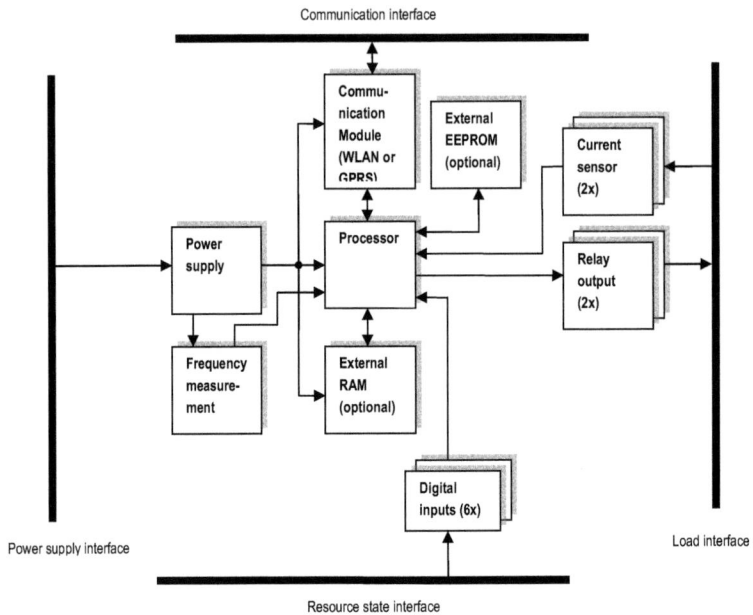

Figure 7.3: Functional blocks of the DSM interface unit

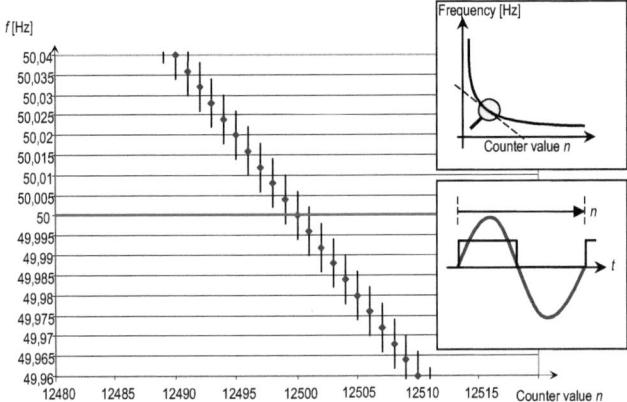

Figure 7.4: Precision of the frequency measurement. The counter result n refers to the duration of the sine oscillation. The indicated error is calculated on the basis of a worst-case crystal error of 120 ppm.

Table 6: Result overview of Internet access technology comparison

Technology	GPRS	Modem DialUp	WLAN
Access technology	GSM/GPRS	Modem ITU-T V.92	IEEE 802.11
Manufacturer	Sony Ericsson	Unbranded	Avisaro
Product name	GR47	-	WLAN module RS232
Communication			
Achievable data throughput (theoretically)	56 kbit/s	56 kbit/s	115 kbit/s
Typical data throughput in tests	9,6 kbit/s[2]	30...40 kbit/s	9,6 kbit/s[3]
Duration of connection initiation	1,4 s	50 s	0,5 s
Typical latency (in tests)[4]	15 ms	210 ms	6 ms
Costs			
Hardware	125 EUR	80 EUR	140 EUR
Connection initiation		-	
Billing base unit	Data volume	connection duration	no additional costs if existing infrastructure used
costs per base unit	30 ct/50 kbyte[5]	2,5 ct/min[6]	
Estimated costs for total unit in high volume	90 EUR	60 EUR	100 EUR
Installation			
Configuration effort needed during installation	Low: Insertion of SIM-Card	High: Connecting telephone terminal, configuration of pre-dial numbers etc.	Low: automatic SSID search and connection attempt possible. Encryption key might be needed. (High if no WLAN available and needs to be set up)

5. The ***interface to the DSM resource*** is realised by a relay (16 A rating) and a current sensor in series, which measures the current drawn by the external device. The current sensor (0...25 A) is a small-footprint hall sensor that transforms current into a proportional output voltage, which

---

[2] Restricted by low-profile processor used in tests (AVR @ 4 MHz)

[3] Restricted by low-profile processor used in tests (AVR @ 4 MHz)

[4] Total latency = Access latency + Internet latency + Server latency. Here, only total latency is given.

[5] Depending on service provider, place and time. Data reflect a "good offer" for prepaid Internet access in Vienna, Austria, in spring 2006. A more recent and comprehensive study of communication costs in the power grid context can be found in [Erb08, p. 16].

[6] Depending on service provider

Implementation

is fed to the microcontroller's A/D converter. With assuming the effective voltage to be 230 V, power measurement can be realised in software. Two relays and sensors are fitted to one interface unit, so that the unit can be used for interfacing two DSM loads. Further, optional additional external RAM and a serial EEPROM device are provided; the latter can be used to store activation levels and activation logs.

6. **Board design:** For hosting the complete circuitry, two 60 x 80 mm double sided boards per unit (one for the 230 V part, one for the low voltage part) were designed and manufactured. The complete schematics are included in Appendix B.

7. **Prototype assembly**: Six prototypes were assembled, three equipped with the WLAN interface card, the other three equipped with the GPRS modem. These are depicted in Figure 7.2.

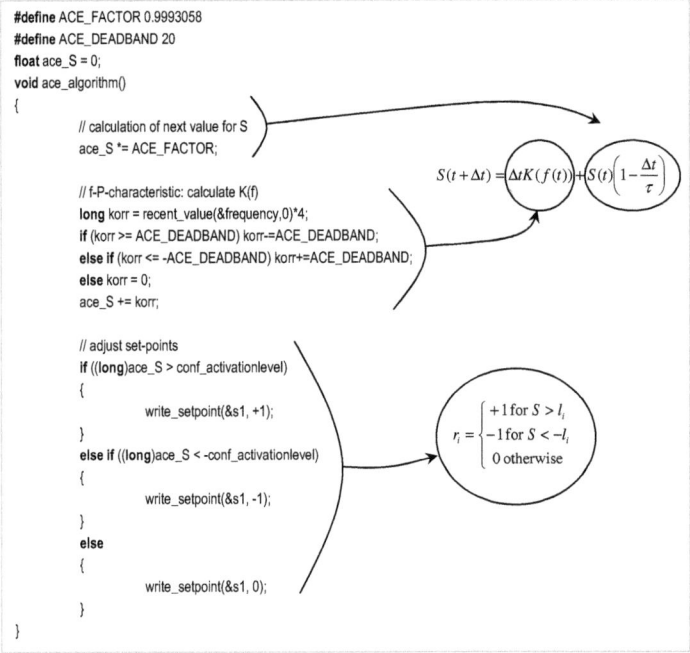

Listing 7.1: Code section of the DSM interface unit software that realises the IRON-CPP algorithm (compare (55) and (68)). The function *ace_algorithm()* is called each time-step ($\Delta t = 1$ s)

8. **Realising remaining functionality in software**: While the basic functionalities of the interface unit are realised in hardware, supervisory functions and the actual DSM algorithm were implemented in software. Since no operating system is used on the microcontroller platform, a simple

non-preemptive task scheduler was implemented, which manages the concurrent jobs executed on the DSM interface unit (which are: reading measurement data, executing DSM algorithm, maintaining server connection, handling server commands). The implemented task scheduler has a fixed turnaround time (the time, in which all tasks are called once) of 1 s. Each task is called every second. It is assumed that the task gives back the control to the scheduler in time and that all tasks together never keep the control longer than 1 s in total. Since there is only a restricted and known set of tasks, the programmer can easily take precautions that this schedule is kept. In case a task does not play after these rules, it is stopped by a watchdog timer. Even though the TCP/IP stack implementation is encapsulated on the communication module and is not realised on the microcontroller, most of the code deals with server communication issues and task management. The actual DSM algorithm is realised in a few lines. The relevant code section is listed and annotated in Listing 7.1. The new value for S is calculated according to (68) on the basis of a current frequency measurement and the function $K(f)$, which is hard-coded in this case. After that, the set-point for the connected thermal process is determined according to (55). In a future version, $K(f)$ should be implemented in such a way that it can be updated during run-time. The shape and scaling of $K(f)$ determines how much control energy is provided by the overall DSM system per mHz frequency deviation. This value should be configurable.

7.3 Central server implementation and communication protocol

According to the DSM algorithm discussed in Chapter 5, the main task of the central server (or the group of central servers) is to calculate and distribute the activation levels regularly. However, with only six real DSM interface units in the field, the calculation of activation levels is a trivial task. In fact, the server of the prototype setting is primarily used for collecting measurement data from the field nodes. Information about network availability of the field nodes, temperature set-points and power consumption of connected electrical loads are requested from the field node and stored in a backend database. The server application is written in Java programming language, the backend database in use is MySQL [8].

The server software listens for connections of field nodes at a specific port and starts a dedicated thread for each connecting unit. Due to restrictions of the communication channel (Internet connection via firewalls, dynamic IPs etc.) the field node has to initiate the connection to the server and not vice versa. During the login process, the field node has the role of a client in the classical client-server paradigm [Tan06 p. 42]. After the login process the field node changes into a mode where it handles commands and requests from the central server. Strictly speaking, the field node takes over the server role. As long as the node is reachable, the central server calls diverse functionalities such as status requests, transmission delay measurements and commands for load control. This process is either terminated by a connection breakdown or by a field node logout.

Implementation

The communication between field nodes and server, which is an application on top of the TCP/IP protocol, is realised using a text-oriented proprietary control protocol (see Appendix C: Communication protocol). While it is also possible to use high-level Internet protocols like XML or SOAP protocols [Ish06], or even embed the DSM commands in HTML code for better relaying by web-optimised services, the choice fell on this particular proprietary solution due to the restricted computational resources on the field node side. Parsing XML code is a challenging task for a microcontroller, especially in terms of memory requirements. A second reason is that a proprietary protocol can be easily designed in such a way that the amount of bytes exchanged between field node and server is as small as possible. This is of interest if a communication channel with volume tariff is used.

7.4 Implementation results

Above, questions have been raised that motivated the practical implementation of the IRON-CPP system. The first question targeted the system complexity: How complex is the infrastructure needed to execute the proposed algorithm? The implementation has shown that a low-cost, microcontroller-based design is sufficient for the DSM interface units. The highest complexity lies actually in the communication interface, which is usually wireless and requires considerable signal processing. In the prototype design, readily manufactured communicaion modules were used.

Regarding the question how the communication between the nodes actually can be realised, it was found that out of the number of potenial solutions the wireless technologies are advantageous in terms of costs as well as installation effort. Nevertheless, a third option that was not explicitly tested is potentially a good choice for the proposed system: Distribution grid operators are gradually setting up so-called smart metering systems, that allow a remote reading of customers energy meters. These systems make use of all previously discussed technologies, including power line communication. In future, a low-bandwidth data connection will be provided to each single household meter by these smart metering networks. Since the alhorithm proposed in this thesis has no real-time requirements for the underlying communication infrastructure, it is perfectly suited to make use of the connectivity provided by these remote metering networks, which will be available for low-cost (both the required hardware, since it is a mass market product, and the communication costs, since the metering system is installed (and financed) for the purpose of meter reading, so DSM is a welcome add-on to the system.

Rearding the costs of the DSM interface unit, these can now be estimated on the basis of the developed prototype. The experimental unit has a size of 75 x 70 x 110 mm and consists of 120 components that account for material costs of 230 Euro. This is for the single unit, components purchased in very low volume (only for 6 prototypes). Still, large potentials for optimization in terms of size, complexity and costs exist. Considering prices for comparable mass-market products, it is anticipated that the device could be manufactured for costs well below 100 Euro [Stad07].

Implementation

The remaining question is: how can actual interfaces between communication node and electrical load look like? For the prototype system, the temperature set-point of thermal processes was changed using the relay contacts of the interface unit. Yet, this was only possible with overruling the device's own controller, which is not a general solution to the problem. The load-side interface is strongly depending on the application context. In a buiding, where the target group of electrical loads are air-conditioning and ventilation systems, load management can be realised over building automation systems such as LON [EN05], BACnet [ISO03]. In an industrial context, again other automation busses and access protocols exist. Generally, future versions of the interface unit will have to be eqipped with a number of physical interfaces and be able to communicate to the electrical load using a variety of different protocols.

The design of the DSM interface unit and the server software has also been tested under field conditions in a waste water treatment plant, where the interface units were used for data collection (Figure 7.5). The implemented prototype system has proven that the proposed system is principally functioning. Yet, as stated before, the implementation and test finally does not give any deeper insight in the behaviour of the algorithm itself due to the small number of resource nodes that have been realised. In this regard, only simulations can be used for testing and optimisation of the algorithm.

For an effective and efficient future realisation of the proposed system for control power provision by DSM, it is however of high importance that the aspects of communication, business processes and hardware integration are strictly standardised. Open standards are the prerequisite for a broad support by different suppliers. It is concluded that it will be necessary in future to integrate the technology, which currently is implemented in the stand-alone DSM interface unit, into the end user equipment itself. Only by this measure, DSM can be operated effective and in a cost-efficient way.

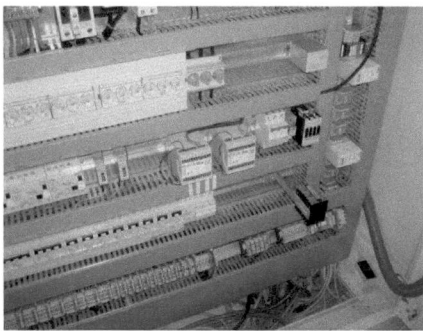

Figure 7.5: Two interface units in action for data measurement in a waste water treatment plant

129

8. Conclusions and Outlook

In this thesis the motivation, design, optimisation and implementation of a distributed algorithm for frequency-responsive provision of control power from electrical loads in the electric power grid has been presented. The key findings are discussed in the following. An outlook in the form of next steps and a vision for the synthesis of computer technology and power engineering in the area of electric power grids is finally presented.

8.1 Main contribution

The questions which motivated this work (see Section 1.4, page 15) can now be answered:

- *How can primary control power be provided by energy consumers?*

    The key element for this is the use of the demand-side storage potential. In many electrical consumption processes, electrical energy is converted into some other kind of energy, such as thermal or potential energy, and this form of energy can be stored in the process due to its inertia for some time. By aggregating a large number of such small demand-side storages, it is possible to modulate the electrical power consumed by these loads without a noticeable loss of energy comfort for the end-user.

- *How can the control algorithm look like?*

    When aggregating a large number of small demand-side storages, so that they appear as one large storage, a control algorithm is needed to coordinate all these small storages. Depending on the required power to/from the aggregated storage, it has to control the individual resources in such a way that the sum of all resource powers results in the required overall power. In the case of primary control power provision, the required power can be calculated from the grid frequency. Thus, each individual resource can act upon this ubiquitous information. The algorithm proposed in this work arranges the individual storages on the scale of energy stored in the apparent large storage. When the apparent storage is required to provide control power, the individual resources discharge one by one in the given order. Simulations results suggest that this approach of a sequential storage utilisation is more robust against synchronisation effects than approaches in which all resources contribute a small fraction at the same time. Synchronisation effects, i.e. multiple loads switching on and off at the same time, can lead to grid instabilities. Sequential storage utilisation however requires communication between the individual resources.

- *What infrastructure is necessary for that?*

The algorithm is designed in such a way that no real-time communication is necessary between the individual resources. In this work, a central (redundant) server is proposed that takes over coordinating tasks such as re-arranging the resources on the storage scale (to provide fair workload arbitration) and also to keep track of (temporarily) unavailable and new resources. For communication, existing wide-area infrastructures such as the Internet or a remote meter reading system can be used.

- *How complex and costly is the hardware needed at the individual electrical loads?*

The complexity of the hardware needed at the individual electrical loads is determined by the communication interfaces, on one side to the electrical load (for storage management) and on the other side to the wide-area communication infrastructure (e.g. GPRS modem). The control algorithm itself is only a small signal processing block. From the implementation done in this work and considering prices for comparable mass-market products, it can be anticipated that the device could be manufactured for costs well below 100 Euro.

8.2 From demand side management to control power provision

The motivation of this work emerged from the context of the upcoming emergency situation in the energy domain, which is determined by the desire for an increased energy-independence of the European Union, the need for CO_2 reduction, and upcoming shortages in fossil resource availability. It is necessary to improve the efficiency of electric energy systems. Energy efficiency measures can be distinguished into two general categories: The first is to reduce the amount of energy wasted in conversion processes. In light bulbs for example, electrical energy is transformed into light energy. However, in low-efficiency light bulbs, a large amount of the electrical energy fed into the bulb is actually transformed into thermal energy, which is unwanted in many cases. This can be reduced by usage of high-efficiency light bulbs. The same principle applies to many other energy conversion processes. The second category is to switch off unused services, such as a light bulb in an unoccupied room. While the first category is a matter of "offline" exchange of equipment, the second category requires decisions and actions during the "run-time" of the energy consumption process. It is the application field of this second category of efficiency measures, where communication and information technology, along with automation infrastructures, play an important role in scenario recognition and automated control of energy processes. When measures are focussing on the *demand side* of the electricity system, they are called demand side management (DSM).

DSM itself is a tool for achieving a certain aim, subsequently referred to as the *DSM application*. At the latest by the liberalisation of energy markets in Europe, the benefit of DSM applications must be measured economically. This thesis bases on the results of previous research, the IRON project [Kup08], a collaborative Austrian project that has studied the economic effects of DSM options in the power grid. Market models for integrating intelligent, on-line DSM solutions into the existing European and especially Austrian power market were studied [Kup08, pp. 43–77]. Closer examinations revealed that only two market models, time-varying energy tariffs and provision of control energy, are economically viable. Especially the control energy model turned out to be very attractive, since control energy is already sold for comparably high prices and a rise of control energy demand can be expected due to the increase of wind energy generation in the grid, making this model even more attractive. By providing control energy from load management, less conventional control capacities have to be allocated and consequently CO_2 emissions are reduced. A short-term lack of generation is not counteracted by an increase of generation from a conventional power plant but by reducing the load. This thesis has dealt with the technical realisation of the DSM application "control energy provision".

In this context, DSM is primarily used to shift consumption times, but not specifically for reducing the energy demand in total. Efficiency improvements are achieved by a better correlation between generation and demand, enabling in the use of more efficient power plants and are also resulting in the reduction of line losses. A reduction of the electricity demand at one time is usually followed by an increased demand at a later time (rebound effect). In contrast to state-of-the-art load shedding, which is performed in grid emergency situations, the DSM algorithm proposed in this thesis is targeted for the normal operation of the grid. Therefore, it has to be taken care that the impact on the energy service which the consumer receives from the grid is not too high. Electrical load management measures have to be performed automatically and hidden. Consequently, only those electrical loads with a certain amount of flexibility in their consumption behaviour are targeted. These are primarily systems where electrical energy is converted into some other form of energy (thermal, potential energy etc.), which then can be stored for some time in thermal capacity of air-conditioned rooms, water heaters and so on. In this work, these demand-side storage systems are called DSM resources.

8.3 Defining factors for the DSM algorithm performance

A generic behavioural model for such DSM resources has been derived (see Chapter 4). It has been found that the load model, together with the DSM application and the underlying communication infrastructure, is a determining component of a DSM system (Figure 8.1). It is specifically determining the algorithm design. The model derived in this work is generic enough to model a large variety of potential DSM resources. For achieving this generality, compromises in terms of model precision have to be made. The model is based on lossless energy storage with restricted capacity.

Conclusions and outlook

In order to match the model to the actual resource behaviour, the lossless energy storage is extended by a linear loss component. Losses are proportional to the amount of energy stored. The model behaviour has been found to be similar to an electrical RC circuit, where the electrical current models the resource power, the voltage on the capacitor is the energy stored in the resource and the resistor models the losses due to non-ideal isolation of the storage. Although the design of this DSM resource model was motivated by the basic physical structure of thermal energy storages, it can be applied to any other kind of demand side energy storage, such as material storage or partial pressure of oxygen in the air of a ventilated room. Also primary energy storages in generators can be modelled with the same approach, which enables to integrate distributed generators into the proposed DSM system.

The third defining factor for a DSM system after the application and the load model is the communication infrastructure. In the current electric power system, communication to the demand side as well as to small generators in the medium and low voltage levels is mostly non-existing. However, for taking influence on the consumption patterns of electrical loads (load management/demand side management) an appropriate communication infrastructure is needed. The costs of this infrastructure, for its setup, its operation as well as for setting incentives for users to join demand response tariffs, must be covered by the advantage gained from the application of the DSM system. In order to keep communication costs as low as possible, existing infrastructures have to be utilised where possible. Since significant distances have to be covered in power systems, Internet communication, mobile communication services or remote meter reading infrastructures are the only available options. Especially the remote meter reading infrastructures, which are currently deployed in European power systems, are a candidate for hosting DSM communication. However, all available low-cost infrastructures allow only low bandwidth communication and do not guaranty any real-time communication properties. This is another defining factor for the design of the DSM algorithm.

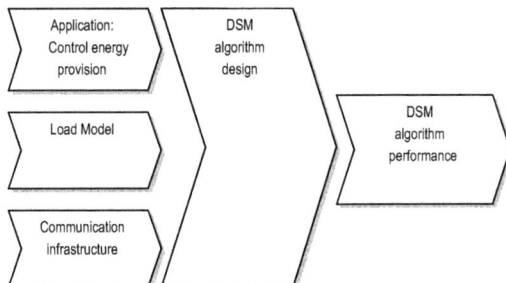

Figure 8.1: Defining factors for the performance of the DSM algorithm are the application, the underlying load model and the properties of the communication infrastructure.

For providing control energy by a pool of DSM resources, this thesis deploys a complete technical concept (see Chapter 5). The algorithm is capable of managing DSM resources in such a way that the modulation of their power consumption is accordance to the requirements for primary control power as defined by [UCTE04a]. The algorithm design is based on the fact that the demand for primary control energy is conceptually broadcasted by the grid frequency, more precisely by its deviation from the nominal value, which can cheaply be measured anywhere in the grid. Therefore, the communication infrastructure has only to fulfil very moderate requirements, since the real-time communication aspect is provided by the grid itself. Internet communication or the emerging infrastructure for smart metering can be used to fulfil these communication demands. The developed "IRON-CPP-PI" algorithm uses frequency measurement in each single DSM node. The frequency information is mathematically transformed into the desired system energy S, which is the sum of energy stored in all DSM resources of the system. The value of S is identical in all nodes (apart from measurement errors and sampling time variation) and serves as a basis for local storage decisions. Each node has a unique activation level. If S is above that, it contributes to the system by charging its local storage. The resulting changes in power consumption of the DSM resources add up to form the desired control power.

The IRON-CPP-PI algorithm developed in this thesis was then tested and benchmarked in simulations (see Chapter 6). Real-environment tests of the proposed system would require a very large number of implemented nodes and thus substantial investments. They also would bear unknown risks to the power system. Therefore, the system can only be evaluated in simulations. Only a small-scale real-world implementation has been realised for testing the algorithmic and communicational aspects of the algorithm (see Chapter 7), while the power-engineering aspects have only been studied in simulations. The simulation results reveal that the proposed IRON-CPP-PI algorithm is performing well and fulfils the requirements. Comparisons to alternative frequency-responsive DSM approaches, such as distributed under-frequency switches or direct and frequency-proportional setpoint variation of thermal processes (discussed in Chapter 3) show that the IRON-CPP-PI achieves in average faster frequency restoration and in all cases better frequency control performance. IRON-CPP-PI achieves a lower integral frequency error in the test cases and delivers better frequency stability. Key factor for the good performance of IRON-CPP-PI is the joined consideration of the DSM application, load model and communication infrastructure for designing the algorithm (Figure 8.1). Only with an appropriate load model, which is also reflected in the DSM algorithm design, it is possible to achieve good results. The comparisons with alternative approaches have confirmed this.

8.4 Steps towards a realisation of control power from the demand side

The current design of the IRON-CPP system, as it is presented in this work, is ready for field test implementation, in which a medium number of nodes take part. Still, the power impact on the system frequency can only be simulated by amplifying the power of the small real resource set. How-

Conclusions and outlook

ever, some aspects remain, in which further research can still be invested. These aspects were not covered in this thesis because they are not its central focus.

The first of these aspects is the clock synchronisation among the resource nodes. These nodes are connected to a central server, which can also take over the roll of a clock master in the system. However, all nodes are also connected to the power grid, which can potentially provide synchronisation signals on the basis of the grid frequency. Either the zero-crossings of the grid voltage itself or distinct events on the frequency function of the grid voltage (over- or under-frequencies in a certain pattern) might be used as synchronisation events. The question is, if the utilisation of such events allows a more precise synchronisation than conventional techniques for client-server systems, depending on the communication infrastructure in use.

The second aspect is the potential use of an alternative system topology. In contrast to the approach discussed so far, which follows the client-server paradigm, also the design of a peer-to-peer algorithm that solves the same task of activation level update might be possible. A number of different definitions for the term "peer-to-peer" exist. However, the most common one, which became accepted in literature, is that peer-to-peer networks are *not* client-server networks [Mahl07, p. 7]. The absence of a central server has the advantage that there is no single point of failure, and the scalability of the system is improved. The current design avoids the single point of failure by a set of redundant central servers (see Section 5.3). However, still the scalability could be improved by a peer-to-peer-oriented scheme. It is now assumed that a peer-to-peer networking approach can also used for the activation level update of the proposed IRON-CPP system. Consequently, no central server can execute the algorithm for activation level update and broadcast the new list to all nodes. The n resource nodes have to coordinate the activation level update among them. This results in a set of special assumptions for the new update algorithm: No node knows all other nodes due to restricted memory resources. Each node holds a *contact list* which keeps record of a few "neighbouring" nodes (which not necessarily have to be physically close). Each node knows its activation level l_i and activation history $a(t)$, and can find out those of its "neighbouring" nodes. Each node knows the current value of S. A global time exists, all resources are synchronised within a certain error interval. Each node can fail and go offline at any time. The task would now be to find an algorithm that updates the activation level in this system in such a way that resources are in average equally loaded.

Apart from finding potentially better solutions to some of the detail aspects of the proposed system, it is essential for an effective and efficient realisation of the proposed DSM technology that the aspects of hardware integration, communication and business processes are strictly standardised. A large step in this direction is done by "Open Automated Demand Response Communication Standards" worked out by the Demand Response Research Center, which is managed by the Lawrence Berkeley National Laboratory [Pie08].

It will be necessary in future to integrate the algorithm, which currently is implemented in the stand-alone DSM interface unit, into the end user equipment itself. Only by this measure, the DSM application becomes throughout economically viable and can be used to change power grids from centralised and passive structures to active grids. In future, a DSM interface can be integrated in every single air conditioning controller, in every heat pump and in every ventilation system. All these interface nodes need to be cost-effective and interoperable. Open standards are the prerequisite for a broad support by different suppliers. Only if control power provision is not a feature of the production line of a single manufacturer of e.g. refrigeration systems, but is supported by multiple manufacturers, the system can gain broad success and the label "DSM ready" becomes a selling argument for appliances.

Standards are required on both sides of the DSM interface, on the communication side as well as on the load interface side. The communication protocol must be supported by all DSM resources and the central servers. Different physical communication media have to be supported to enable a broad deployment. Power line carrier (widely used by remote metering systems on the "last mile" in the low voltage grid level) and different wireless technologies will have to be utilised. On the other side of the interface, the modulation of the power consumption of the electric loads is usually only possible when accessing internal control signals. This technical barrier can only efficiently be skipped by integration of the DSM node into the end-user appliance, so that the manufacturer can make use of his specialised knowledge of the internal mechanisms of the appliance.

Not only the technical aspects have to be solved efficiently, also the installation and operation of the DSM system components must be solved economically. The system must be "plug & play", meaning that the DSM nodes connect to the DSM infrastructure without complicating configurations by the installing person. The node should configure itself and announce its properties to the infrastructure.

8.5 Information technology in the power grid: the paradigm shift

DSM is one example which illustrates the need for information and communication technology in the power grid. This thesis demonstrates how in detail the task of these technologies in the context of energy provision can look like. DSM is only one among a variety of cases, where information technology can improve or even revolutionise aspects and solutions in the power grid and energy systems in general.

The electric transmission and distribution grid, which was designed as a passive grid infrastructure, is facing a substantial growth in the number of sensors, which are remotely interpreted and measure on-line data from the grid such as voltage, current, frequency, power, energy, flicker, harmonics and other data. In the past the grid operator was virtually "blind" on the medium and low voltage level regarding on-line information of power flows. Measurements usually were and are done by local drag indicators, which have to be read on-site, and only inform about maximum, minimum or mean values of the measurement period. However, in recent years a number of drivers have emerged that result in more and more measurement systems to be installed and read remotely.

Conclusions and outlook

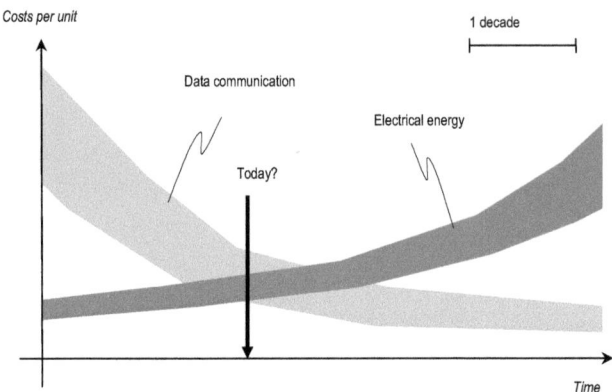

Figure 8.2: Estimated development of costs for communication and energy

One of these drivers is the advance in information and communication technologies. The availability of inexpensive large memory capacities allows for storing data series instead of maximum, minimum or meaning values. Innovations in communication systems, especially in the areas of signal processing and in production technologies, have resulted in the deployment of communication systems that enable comparably high data throughput for low costs. Wireless transmission of data is state of the art at this stage for remote control in medium voltage grids, a fact that shows that this technology has reached an adequate level of maturity and is accepted by the grid operators, who traditionally are very concerned about the reliability of information technology in the grid. The technological advance on the side of information technology on one hand and the beginning shortage in energy supply (including the need for CO_2 reduction) on the other hand result in an economic paradigm shift: as costs for energy rise and costs for communication fall, the relation between both begin to change and the unit cost for energy is ultimately becoming higher than the unit cost for communication in many application areas (Figure 8.2).

Another driver is the integration of distributed generators, especially from renewable energy sources, into the existing power grid. For a sustainable and secure future electrical energy provision, it is inevitable to significantly increase the share of generation from renewable sources in the energy system. Compared to the traditional generation from fossil resources, the energy density of renewable energy sources is low. The number of generation units is comparably high, but they have a rather low individual power output compared to large centralised power plants. The integration of such distributed generators into the existing power grids leads to a number of different issues [Lug07]. One of these issues is the fact that a strong growth of electricity generation in the medium voltage grid, where most of the installed distributed generation injects its power, leads to grid voltage problems [Jen00]. At the feed-in points, the grid voltages reach the given limits in times of low demand, so that no more units can be installed without significant grid investments [Kup07a]. Here, on-line voltage control in the medium voltage feeders (basing on measurement data from critical grid nodes) by controlling grid components such as tab-changer transformers and generation units, so-called active grid operation [Ham06], can solve the problem of keeping the grid voltage in the defined limits. Instead of investing into new power lines, the problem is tackled by installing comparably inexpensive communication systems and controllers [Pru08]. This is an example for an application area where the intersection point of communication and energy costs as shown in Figure 8.2 has already been passed.

It can be expected that energy system will in future be operated closer to their limits as it is currently the case. One of the reasons for that is that the pattern and kind of investments into the grid infrastructure will change due to the liberalisation of power markets. For maintaining the high standards in power quality, it is already today considered to be necessary to monitor power quality variables such as voltage, flicker and harmonics using on-line measurements in the grid. This is another driving factor for an increasing flood of on-line measurement data from the grid.

Taking all these driving factors together, it is foreseeable that information from the grid will become more than any conventional SCADA (supervisory control and data acquisition) system can handle. For tackling this emerging complexity, it is necessary that the information handling is done in a much more decentralised manner than it is done today. The development from large and bulky centralised systems to lightweight distributed solutions can also be seen in other areas, for example in the field of building automation. Here, the transformation from central computers which are connected to each sensor and actuator in the building over a dedicated wire towards distributed control networks has already taken place. Modern building control systems utilise communication busses and distribute the computational power over the individual network nodes [EN05, ISO03]. Even in such a safety-critical application area as aircraft control, information technology and communication systems are common today [Wol08]. This example shows that it is possible to design highly dependable and fault-tolerant control systems, which ultimately improve the safety of the application system (here: aircraft) instead of having a negative influence on dependability. The fear that information and communication technology will have negative impact on dependability the power grid as a whole is still common among grid operators today. It mainly originates from the false impression that the level of dependability achievable by modern information systems is equal to the dependability we experience from consumer products on a daily basis.

8.6 Vision of smart electricity grids

As outlined in the previous section, strong drivers are working towards more information and communication technology in the power grids. Of course, information technology is not an end in itself in this context, but a means for more efficient, sustainable and cost-effective provision of electrical energy and ancillary (i.e. supporting) services. The vision of the future power grid with an increased

Conclusions and outlook

level in utilisation of information technology is that of a "smart grid". The term "smart grid" is currently promoted by the European technology platform of the same name [9], which is formed by diverse stake-holders in the field. It is aiming to define a realistic vision of such intelligent power networks and to harmonise research and development work on European level in order to come closer to the vision. Also on national level, a similar technology platform is in the process to be installed. The first task of the "National Technology Platform Smart Grids Austria" will be to deliver a definition of smart grids for Austria. In the following, the author's point of view regarding the meaning of smart electricity grids is outlined.

Control measures on the demand side play a role in most "smart grid" visions. They are seen as a supporting tool to match supply and demand under the condition of supply from fluctuating renewable energy resources, whose generation patterns do not match the demand curves. It however also makes sense to adjust the generation to the supply by featuring generation technologies where this is possible, e.g. for residential combined heat and power (CHP) systems. Here, energy can be stored on the primary energy side as well as on the thermal side. However, it makes sense to combine the operation of multiple small CHP units as well as DSM measures to gain the advantage of a high number of freedoms in the system. Comprehensive control of distributed energy resources, either generators or loads, will be necessary. For this, information exchange between the grid participants is needed. The future smart grid will be characterised by an intensified flow of information compared to the state-of-the-art power grid, where the dominating flow of energy is only accompanied by sporadic (monthly or yearly) meter readings.

Although the term smart or intelligent network, which origins from the area of telecommunication systems, is more and more often used in the context of electricity grids (see [Cont06], [Lay02] or [9]), there is no common definition for what makes an electricity grid actually intelligent. To separate the term from its telecommunications counterpart, the term "intelligent grid" will be used here. In a strict technical context, the term *intelligent* refers to properties such as autonomy, context-awareness, to the ability to learn, make use of gathered knowledge (data) and the ability solve problems. (This understanding of technical intelligence, if such a form of intelligence exists at all, is much more comprehensive than the understanding implied by the widely adopted trend to call any new technical solution "intelligent".) Admittedly, an *intelligent grid* will not factually behave intelligently, but its components will feature functionalities that emulate intelligent behaviour.

Many large network systems tend to develop, as they grow, towards huge hierarchical structures with relatively simple components on the lower levels. This can be observed e.g. with telecommunication systems or the Internet, but also with electricity grids. Networks of this kind are able to meet growing demands, such as rising data communication bandwidth or rising energy consumption, but it is very hard to implement additional features or structural changes into these "large and dump" networks. At this point, the idea of intelligent networks becomes eminent.

As outlined before, European electricity grids are facing such sharp and rapid structural changes. The implementation of upcoming requirements will involve innovative technical solutions that significantly differ from standard measures for grid update. Electricity grid investments are done for long-term time horizons. Thus, new components should to be designed in such a way that future technical demands, which are not predictable today, can be fulfilled. A brute force approach for this would be over-dimensioning, but this is economically unfeasible.

The key for this upcoming design decision is the intelligent grid. Autonomously acting components stay operative even in grid failure situations; their context-awareness enables them to adjust to different situations. Due to the ability to learn from collected data, the grid will be able to solve problems locally if possible and even handle new, previously unknown situations.

The challenge will be to design the components in such a way that on one hand they behave in the sense of technical intelligence and act autonomously, but on the other hand their actions and reactions are comprehensible, logical and traceable. The proposals in this thesis are a first step towards this smart grid vision, since it will feature the use of more communication and information technology in the grid, which is a prerequisite for future "smart grids".

Appendix A: Resource sets

Resource sets are used for the simulation of the proposed algorithm for control power provision by DSM resources. In this thesis, three different resource sets are used. All resource sets base on the measurements performed on a real DSM resource (see Section 4.3.3) and have a set size of $N = 1000$. The difference between the resource sets is the randomising deviation. The parameters and values of the resource sets are shown in Table A.1, histograms of the time constants for the resource sets in Figure A.1 ... A.3.

Table A.1: Resource Sets A, B and C

Variable Name	Symbol used in Formulas	Value	Deviation			Description
			Resource Set A	Resource Set B	Resource Set C	
Pel_on	$P_{el,on}$	150 W	+/− 20 %	+/− 50 %	+/− 10 %	Power consumed by the resource when thermostat is on on-stat (0 otherwise)
eta	η	0.8	+/− 4 %	+/− 10 %	+/− 2 %	Efficiency of the transformation of electrical onto thermal energy
R	R_L	0.5 K/W	+/− 20 %	+/− 50 %	+/− 10 %	Thermal resistance of the isolation
C	C	5000 J/K	+/− 20 %	+/− 50 %	+/− 10 %	Thermal capacity
Tset	T_{set}	18	+/− 20 %	+/− 50 %	+/− 10 %	Temperature set-point
Tdiv	$\Delta T_1 = T_{set} + \text{Tdiv}$ $\Delta T_2 = T_{set} - \text{Tdiv}$	1	+/− 20 %	+/− 50 %	+/− 10 %	Temperature deviation allowed by the thermostat (positive and negative)
Tshift	ΔT_{set}	1 °C	+/− 20 %	+/− 50 %	+/− 10 %	Set-point shift for DSM operation
num_resources	N	1000	+/− 20 %	+/− 50 %	+/− 10 %	Number of resources in set

Appendix

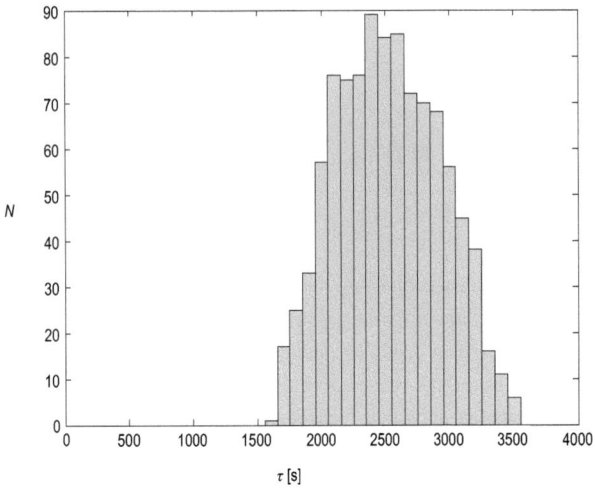

Figure A.1: Distribution of τ for Resource Set A

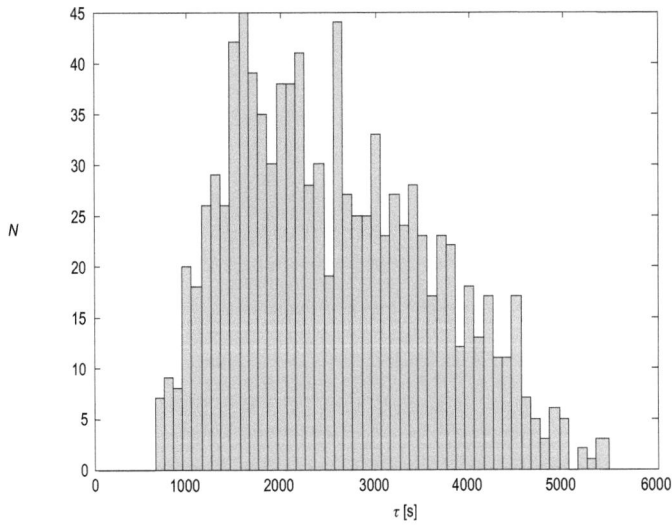

Figure A.2: Distribution of τ for Resource Set B

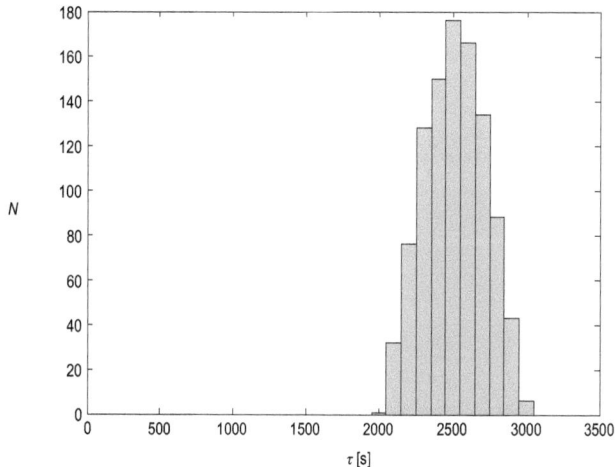

Figure A.2: Distribution of τ for Resource Set C

Appendix

Appendix B: DSM interface unit schematics

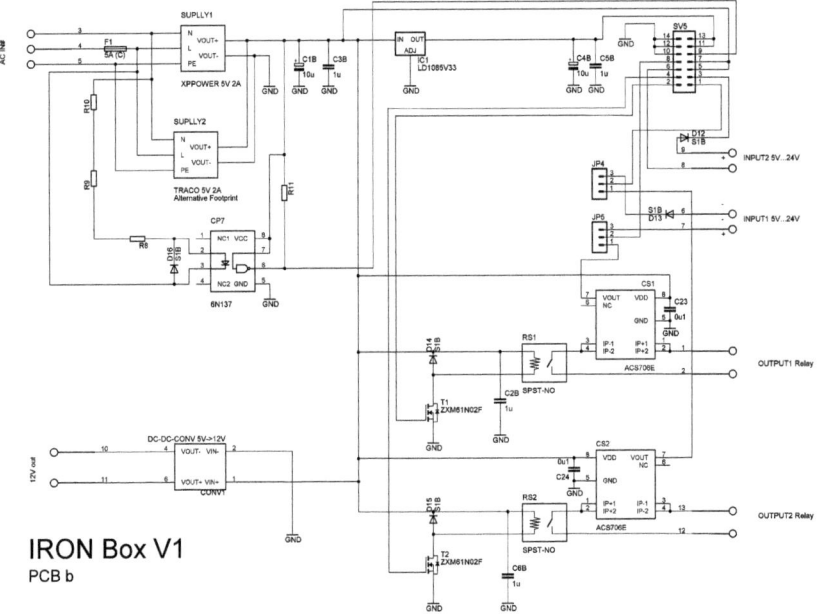

Appendix

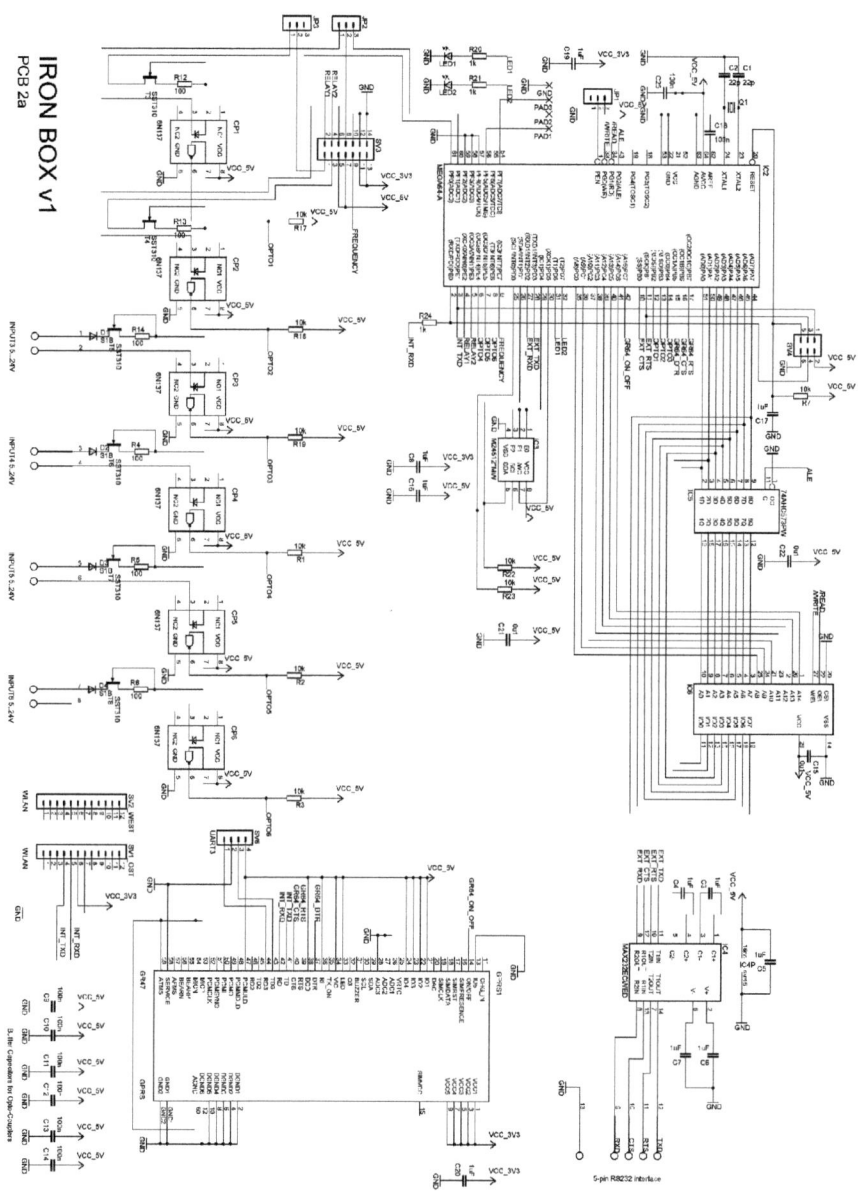

147

Appendix

Appendix C: Communication protocol

IRON Network structure and communication scenarios

Fig. 1 depicts an abstract view of the IRON network structure. The communication links are realised using different communication media, especially GPRS, WLAN and PLC.

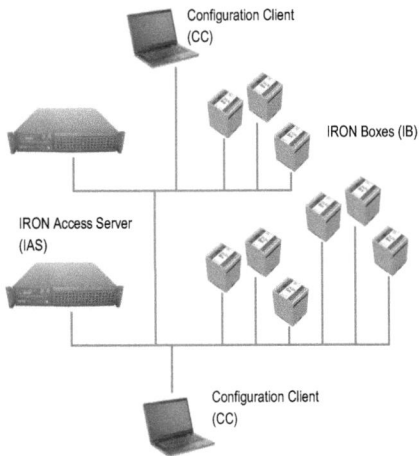

Fig. 1: IRON Network structure

A large number of *IRON Boxes* (IB) is distributed among households, small businesses and industry setups. In the extreme case, every household of a certain area is connected to the IRON system. Each IRON Box can have influence on the consumption patterns of one or more loads. Each box is able to connect to an IRON Access Server (IAS). The selection of the IAS to connect to depends on connection quality and load balancing strategies.

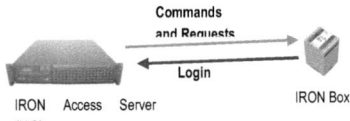

Fig. 2: After the IRON Box has connected to the IAS, it switches to server mode and handles commands and requests from the IAS.

Due to restrictions of the communication channel (Internet connection via firewalls, dynamic IPs etc.) the IRON Box has to initiate the connection to the IAS and not vice versa. During this login process, the Box has the role of a *client*. After the login process the box changes into a *server* mode

Appendix

where it handles commands and requests from the IAS. As long as the box is reachable from the IAS, the IAS calls diverse functionalities such as status requests, transmission delay measurements and commands for load control. This process is either terminated by a connection breakdown or by a box logout.

IRON Protocol history

New commands in Version 0.1 (27.11.2006)

#measurement
#iam
#thatsit
#read

New commands in Version 0.2 (19.12.2006)

#storage
#hello
#time
#setpoint
#status

IRON Protocol details

On the subsequent pages, all IRON Protocol commands are listed in detail.

TU VIENNA — Vienna University of Technology, Institute of Computer Technology	IRON Protocol	Version 0.2
Description of: Format of IRON commands	**Format of IRON commands**	
Syntax #comname argumentlist newline		**Example:** #measurement Temp1, Temperature, K, 1, 0.1, 273, 10, 1000

Description:

Each command is a text string that starts with a hash character (#) and is terminated by the two characters (ASCII 10, ASICII 13).

The maximum length for the command name is 20 characters.

The argument list is a comma-seperated list of argument strings. These can have three different types:

STRING: a string of characters. No commas allowed. Maximum length is 20 characters.

INTEGER1, INTEGER2, INTEGER4, INTEGER8: The representation of an integer value. May have a trailing "-". Apart from that, only digits are allowed. The trailing number *n* in INTEGER*n* specifies the number of bytes used to represent the value. This does not effect the textual representation of the values in protocol transmission, but the allowed value range.

FLOAT: The representation of a floating point number. May have a trailing "-". The fraction is separated by a point ("."). Apart from that, only digits are allowed.

All commands and parameter Strings in the IRON protocol are case sensitive.

Possible answers:

Answers and answer formats are specified for each command separately. In general, each answer is exactly one string, terminated by the two characters (ASCII 10, ASICII 13).

Data items in the answer string can have the same three types STRING, INTEGER and FLOAT as defined above.

Unknown commands cause the answer #UCMD.

All commands and parameter Strings in the IRON protocol are case sensitive.

TU VIENNA VIENNA UNIVERSITY OF TECHNOLOGY INSTITUTE OF COMPUTER TECHNOLOGY	**IRON Protocol**		Version 0.2
Description of: Protocol command Box → IAS	Name: **#iam**		
Syntax: #iam *id*		Example: #iam 00346	

Description:

This command is used by the box after the connection to the IAS is established to inform the IAS about its presence and name.

The parameter *id* is an INTEGER8.

Possible answers:
#OK or #ERROR if the parameter could not be interpreted correctly.

Appendix

TU VIENNA — Vienna University of Technology, Institute of Computer Technology		IRON Protocol	Version 0.2
Description of: Protocol command Box → IAS		Name: **#storage**	
Syntax: #storage: *name, type, capacity, chargetime, dischargetime*			Example: #storage fridge1, A, 50, 120, 278 #storage aircon, B, 2500, 360, 1453

Description:
The box uses the #storage command to inform the IAS about the real or virtual storages connected to the box. For each storage, the command is issued once. Each storage has a *name* (STRING) that is unique among the storage names of this box. Each storage has also a *type* (STRING). Possible types are listed below. The parameter *capacity* (INTEGER4) is the storage capacity in Wh. *Chargetime* (INTEGER4) and *dischargetime* (INTEGER4) are the times in seconds it takes to charge or respectively to discharge the storage.

Possible storage types:

A – Setpoint variation storage. The storage has three energy levels: normal, low = normal – capacity and high = normal + capacity. The IAS is free to set the storage in any of these states.

B – Immediate recharge storage. The storage has two energy levels: normal and low = normal – capacity. It usually is in normal state but can be transferred into low state by the IAS (power supply switched off). The storage itself changes back to normal state (and recharges) as soon as the internal energy level becomes too low.

Note: In a future protocol version, a #restriction command will be present to communicate storage scheduling restrictions to the IAS.

Possible answers:
#OK or #ERROR if the parameters could not be interpreted correctly.

TU VIENNA — Vienna University of Technology, Institute of Computer Technology	**IRON Protocol**	Version 0.2

Description of:	Name:
Protocol command Box → IAS	**#measurement**

Syntax:	Example:
#measurement: *name, type, unit, bytesPerSample, lsbUnit, offset, interval, capacity*	#measurement Temp1, Temperature, K, 1, 0.1, 273, 10, 1000

Description:

This command is used by the box after the connection to the IAS is established to inform the IAS about measurement values that can be read from the box.

Parameters are comma-separated and can be of type STRING, INTEGER or FLOAT:

name: the name of the measurement value. Used for identification. Should be unique for each measured variable. STRING

type: the type of the variable, which can be: "Power", "Current", or "Temperature". STRING

unit: the unit in which the variable value is measured. E.g. "K" or "mA". STRING

bytesPerSample: the number of bytes used to represent a single sample of the variable (applies when measurement data is send to the IAS) INTEGER1

lsbUnit: the precision of the measurement. FLOAT
offset: the offset of the measurement. FLOAT

E.g. if unit is "mA", lsbUnit is 10 and offset is 5, then a transmitted value of 2 means actually 25 mA.

interval: the time between two sample measurements in seconds. INTEGER2

capacity: the number of samples for this variable the box is able to store. INTEGER4

Possible answers:
#OK or #ERROR if the parameters could not be interpreted correctly.

Appendix

TU VIENNA — Vienna University of Technology, Institute of Computer Technology	**IRON Protocol**	Version 0.2

Description of:	Name:
Protocol command Box → IAS	**#thatsit**

Syntax:	Example:
#thatsit	#thatsit

Description:

This command is used by the box after the connection to the IAS is established and it has informed the IAS about its name (using #iam) and its 0…n measurement variables (using #measurement) as well as its 0…m storages (using #storage). It terminates the initial communication with the IAS. After the #thatsit command is sent, the box listens for commands for the IAS.

Possible answers:

#OK or #ERROR

Appendix

	VIENNA UNIVERSITY OF TECHNOLOGY INSTITUTE OF COMPUTER TECHNOLOGY	IRON Protocol	Version 0.2
Description of: Protocol command IAS → Box	Name: **#read**		
Syntax: #read *varname, numOfSamples*		Examples: #read Temp1, 100 #read Power2, all	

Description:

The IAS requests the most recent *numOfSamples* samples of the measured variable *varname*. *numOfSamples* is INTEGER4 in this case. Alternatively, *numOfSamples* can be the STRING "all". In this case, all available samples are requested.

Possible answers:

The samples are transmitted as comma-separated INTEGER4 in order of their age (oldest first). The first value transmitted is the timestamp (INTEGER8) for the oldest sample.

The timestamp counts the number of seconds from a starting point, which for evaluation purposes is the box power up time. This should be updated as soon as possible to an absolute time value.

If not enough samples are available or *varname* is not available, #ERROR is returned. #ERROR can be followed by an additional textual description in the same line.

Appendix

TU VIENNA — Vienna University of Technology, Institute of Computer Technology	**IRON Protocol**	Version 0.2

Description of:	Name:
Protocol command IAS → Box	**#hello**

Syntax:	Examples:
#hello	#hello

Description:

The IAS uses the #hello command to find out whether a box is still on line. Additionally it is used to determine the data transmission time which is used for clock synchronisation (see #time command).

Possible answers:

The box shall answer #OK as quick as possible after receiving a #hello request.

VIENNA UNIVERSITY OF TECHNOLOGY — INSTITUTE OF COMPUTER TECHNOLOGY	IRON Protocol	Version 0.2

Description of:	Name:
Protocol command IAS → Box	**#time**

Syntax:	Examples:
#time *milliseconds*	#time 2300239432

Description:

The IAS informs the box about the current time using this command. The parameter *milliseconds* is INTEGER8 and counts the number of milliseconds from 1.1.1970 0:00:00:00.

The box should react on this command by synchronising in the given time. The IAS takes care that the time value is approximately correct when it arrives at the box by using the #hello command to determine the transmission delay.

Possible answers:

#OK or #ERROR. #ERROR can be followed by an additional textual description in the same line.

Appendix

	IRON Protocol	Version 0.2
VIENNA UNIVERSITY OF TECHNOLOGY — INSTITUTE OF COMPUTER TECHNOLOGY		

Description of:	Name:	
Protocol command IAS → Box	**#setpoint**	

Syntax:	Examples:
#setpoint *storage, status*	#setpoint fridge1, 1 #setpoint ballroom -1 #setpoint office_66 0

Description:
The #setpoint command sets the setpoint for a given storage. The parameter *storage* (STRING) identifies the storage (see also command #storage). The parameter *status* (INTEGER1) identifies the storage setpoint level relatively to the normal state, which is referred to as 0.

For type **A** storages, *status* can be -1, 0 or 1. Negative values result in energy to be released, positive values result in energy to be stored.

For type **B** storages, *status* can only be -1. Type B storages can only have energy levels low and normal; and the transition from low to normal is done automatically.

For more information about storage types see command #storage.

Possible answers:
#OK or #ERROR. #ERROR can be followed by an additional textual description in the same line.

Appendix

TU VIENNA — VIENNA UNIVERSITY OF TECHNOLOGY, INSTITUTE OF COMPUTER TECHNOLOGY	**IRON Protocol**	Version 0.2

Description of:	Name:
Protocol command IAS → Box	**#status**

Syntax:	Examples:
#status *storage*	# status fridge2 # status ballroom

Description:

The IAS can use the #status command to gain information about the current energy level of a storage. For more information about storage types see command #storage. See also command #setpoint for energy levels.

The parameter *storage* (STRING) identifies the storage.

Possible answers:

status, since
or
#ERROR optionally followed by an additional textual description in the same line.

If no error occurs, the box sends back two comma-separated values, of which the first (*status*, INTEGER1) is the current storage status (-1, 0, 1) and the second (*since*, INTEGER8) is the timestamp of when this status became active. If no status change as occurred since the last box reset, *since* shall read as 0.

For type **A** storages, the status should read as the value set by the #setpoint command.
For type **B** storages, the #status command can be used by the IAS to find out (by polling the box) for how long the storage level 'low' was active and when the storage changed to level 'normal'.

References on scientific publications

[Acht04]	Achterberg, T.: SCIP - a framework to integrate Constraint and Mixed Integer Programming, ZIB report 04-19, Berlin 2004
[Awer97]	Awerbuch, S., Preston, A.: The Virtual Utility: Accounting, Technology & Competitive Aspects of the Emerging Industry, Kluwer Academic Publishers, Massachusetts, USA, 1997
[Ban90]	Banerjee, S., Chatterjee, J. K., Tripathy, S. C.: Applictaion of Magnetic Energy Storage Unit as Load-Frequency Stabilizer, IEEE Transactions on Energy Conversion, Vol. 5, No. 1, March 1990, pp. 46–51
[Bar04]	Barbose, G., Goldman, C., Neenan B.: A Survey of Utility Experience with Real Time Pricing, Lawrence Berkeley National Laboratory, Paper LBNL-54238, 2004
[Bla05]	Black, M., Silva, V., Strbac, G.: The role of storage in integrating wind energy, International Conference on Future Power Systems, November 2005, ISBN: 90-78205-02-4
[Bra06]	Brauner, G. et al.: Verbraucher als virtuelles Kraftwerk, Berichte aus Energie- und Umweltforschung 44/2006, BmVIT, Vienna, 2006
[Casaz04]	Casazza, J., Delea, F.: Understanding Electric Power Systems, IEEE Press, 2004, ISBN 0471446521, pp. 1–10
[Casso04]	Casson, H. N.: The History of the Telephone, Kessingner Publishing, 2004, ISBN 1419166, pp. 1–5
[Cont06]	Conti, J. P.: Let the Grid do the Thinking, *IET Power Engineer,* April/May 2006, pp. 34–38
[EN05]	European Norm. EN 14908 – Open data communication in building automation, controlsand building management – control network protocol. CEN, 2005
[Erb08]	Erben, S.: Methoden zur kommunikationstechnischen Integration dezentraler Energieerzeuger in den aktiven Netzbetrieb, Bachelor Thesis (in German language only), TU Vienna, 2003
[Fuenf00]	Fünfgeld, C., Tiedemann, R.: Anwendung der repräsentativen VDEW-Lastprofile, Vereinigung Deutscher Elektrizitatswerke -VDEW- e.V., M05/2000, Frankfurt, 2000 (in German language)
[Ham06]	Hammons, T. J.: Integrating Renewable Energy Sources into European Grids, Proceedings of the 41st International Universities Power Engineering Conference, UPEC '06, Newcastle upon Tyne, September 2006, pp. 142–151
[Ham07]	Hammerstrom, D. J. et al.: Pacific Northwest GridWise[TM] Testbed Demonstration Projcts, Part II, GridFriendly[TM] Appliance Project, Pacific Northwest National Laboratory, Project Report 2007
[Heb02]	Hebner, R., Beno, J., Walls, A.: Flywheel batteries come around again, IEEE spectrum, Volume 39, Issue 4, pp. 46 to 51, April 2002
[IEA03]	International Energy Agency, Cool Appliances, Policy Strategies for Energy Efficient Homes, OECD/IEA, Paris, 2003 (available online, accessed 05/2008: *http://www.iea.org/textbase/nppdf/free/2000/cool_appliance2003.pdf*)
[Inf07]	Infield, D. G., Short, J., Horne, C., Freris L. L.: Potential for Domestic Dynamic Demand-Side Management in the UK, IEEE Power Engineering Society General Meeting, June 2007
[Ish06]	Ishida, S., Roesener, C., Ichimura, J., Nishi, H.: Implementation of Internet based Demand Controlling System, Proc. of 4th IEEE Int. Conference in Industrial Informatics, Singapore, August 2006
[ISO03]	International Organization of Standardization. ISO 16485-5 – Building automation and control systems (BACS) – Part 5: Data communication protocol, ISO 2003

161

References

[Jen00] Jenkins N., Allan R., Crossley P., Kirschen D., Strbac G.: Embedded Generation, The Institution of Electrical Engineers, ISBN 0 85296 774 8, 2000

[Kil06] Kiliccote, S., Piette, M. A., Hansen, D.: Advanced Controls and Communications for Demand Response and Energy Efficiency in Commercial Buildings, Proc. of Second Carnegie Mellon Conference in Electric Power Systems, Pittsburgh PA. LBNL-59337, January 2006.

[Kirby03] Kirby, B. J.: Spinning Reserve from Responisve Loads, Oak Ridge National Laboratory, March 2003

[Koch04] Koch, T.: Rapid Mathematical Programming, PhD Thesis, Technische Universität Berlin, 2004.

[Kund94] Kundur, P.: Power System Stability and Control. New York: McGrawHill, 1994, pp. 581–592

[Kup06] Kupzog F.: Self-controlled Exploitation of Energy Cost saving Potentials by Implementing Distributed Demand Side Management, 4th International IEEE Conference on Industrial Informatics (INDIN 2006), Singapore, September 2006

[Kup07a] Kupzog F., Brunner H., Prüggler W., Pfajfar T., Lugmaier A.: DG DemoNet-Concept – A new Algorithm for active Distribution Grid Operation facilitating high DG penetration, 5th International IEEE Conference on Industrial Informatics (INDIN 2007), Vienna, Austria, July 2007

[Kup07b] Kupzog, F., Palensky, P.: Wide-Area Control System for Balance-Energy Provision by Energy Consumers, proc. of 7th IFAC International Conference on Fieldbuses & Networks in Industrial & Embedded Systems (FeT 2007), Toulouse, France, Nov. 2007, pp. 337–345

[Kup07c] Kupzog F., Roesener C.: A closer look on load management, 5th International IEEE Conference on Industrial Informatics (INDIN 2007), Vienna, Austria, July 2007

[Kup08] Kupzog, F. et al.: Integral Resource Optimization Network – Concept, Project report Energiesysteme der Zukunft, Project No. 810676, BmVIT, 2008

[Kyoto98] Kyoto Protocol to the United Nations Framework Convention on Climate Change, United Nations, 1998 (available online: http://unfccc.int/resource/docs/convkp/kpeng.pdf)

[Law03] Lawrence, D.J., Neenan, B.F.: The status of demand response in New York, Power Engineering Society General Meeting, 2003, IEEE, Volume 4, pp 13–17, July 2003

[Lay02] Laycock, W. J.: Intelligent Networks, Power Engineering Journal, February 2002, pp. 25–29

[Laz04] Lazarewicz, M. L., Rojas, A.: Grid Frequency Regulation by Recycling Electrical Energy in Flywheels, IEEE Power Engineering Society General Meeting 2004, Denver, June 2004, pp. 2038–2042

[Li03] Li, Y., Flynn, P. C.: Deregulated Power Prices: Changes Over Time, IEEE Transactions on Power Systems, Volume 20, No. 2, May 2005, pp. 565 – 572

[Lug07] Lugmaier A., Brunner A., Bletterie B., Kupzog F., Abart A.: Intelligent Distribution Grids in respect of a growing share of Distributed Generation, 19th International Conference on Electricity Distribution (CIRED 2007), Vienna, Austria, May 2007

[Mahl07] Mahlmann, P., Schindelauer, C.: Peer-to-Peer Netzwerke, Springer, 2007

[Nae06] Naedele M.: An Access Control Protocol for Embedded Devices, 4th IEEE International Conference on Industrial Informatics (INDIN06), Singapore, August 2006

[Nie06] Nieuwenhout, F. D. J., Hommelberg, M. P. F., Schaefler, G. J., Kester J. C. P., Visscher, K.: Feasibility of distributed electrictity storage, International Journal of Distributed Energy Resources, Volume 2 Number 4, October – December 2006, pp. 307–323

[Ohm06] Ohm, J.-R., Lücke, H. D.: Signalübertragung, ISBN 3540222073, Springer, 2006 (in German language)

[Pal01] Palenksy, P.: Distributed Reactive Energy Management, Doctoral Thesis, TU Vienna, 2001

[Pal06] Palensky, P. et al.: Integral Resource Optimization Network – Study, Project report Energiesysteme der Zukunft, Project No. 808570, BmVIT, 2006

[Pea07] Pearmine, R., Song, Y. H., Chebbo, A.: Experiences in modelling the performance of generating plant for frequency response studies on the British transmission grid, Electric Power Systems Research 77, 2007, Elsevier, pp. 1575–1584

[Pie08] Piette, M. A. et al.: Open Automated Demand Response Communication Standards, Ernest Orlando Lawrence Berkeley National Laboratory, Draft Version, Berkeley, May 2008 (available online: http://drrc.lbl.gov/openadr/, accessed 13.06.2008)

[Pru08] Prueggler, W., Kupzog, F., Brunner, H., Bletterie, B.: Active Grid Integration of Distributed Generation utilizing existing infrastructure more efficiently - an Austrian case study, Proceedings of the 5th International Conference on the European Electricity Market, Lisbon, May 2008

[Roe05] Roesener, C., Palensky P., Weihs M., Lorenz, B., Stadler M.: Integral Resource Optimization Network - a new solution on power markets, Proceedings of 3rd International IEEE Conference on Industrial Informatics; Perth, Australia, August 2005

[Rup07] Ruprecht, A., Göde, D.: Die Rolle der Wasserkraft im künftigen Strommarkt - Aufgaben und technische Konsequenzen, 5. Internationale Energiewirtschaftstagung IEWT 2007, Vienna, Austria 2007 (in German language)

[Sch05] Schreiber, M., Whitehead, A. H., Harrer, M., Moser, R.: The vanadium redox battery - an energy reservoir for stand-alone ITS applications along motor and expressways, Proceedings of the 8th International IEEE Conference on Intelligent Transportation Systems, pp. 391 – 395, Vienna, Austria 2005

[Shok05] Shokooh, F., Dai, J.J., Shokooh, S., Taster, J., Castro, H., Khandelwal, T., Donner, G.: An intelligent load shedding (ILS) system application in a large industrial facility, Industry Applications Conference, 2005. Fortieth IAS Annual Meeting. Conference Record of the IEEE, 2005.

[Short07] Short, J. A., Infield, D. G., Freris L. L.: Stabilization of Grid Frequency Through Dynamic Demand Control, IEEE Transactions on Power Systems, Vol. 22, No. 3, August 2007

[Smi94] Smith, H. L.: DA/DSM directions. An overview of distribution automation and demand-side management with implications of future trends, Computer Applications in Power, IEEE Volume 7, Issue 4, pp. 23–25, October 1994

[Str02] Strbac, G.: Quantifying the system costs of additional renewables in 2020, Manchester Centre of Electrical Energy, UMIST, Manchester, U.K., Tech. Rep., October 2002, Report to the U.K. Department of Trade and Industry (available online: www.berr.gov.uk/files/file21352.pdf)

[Stad03a] Stadler, I. Bukvic-Schäfer, A. S.: Demand side management as a solution for the balancing problem of distributed generation with high penetration of renewable energy sources, International Journal of Sustainable Energy, Volume 23, Issue 4 December 2003, pp. 157–167

[Stad03b] Stadler, M.: The relevance of demand-side-measures and elastic demand curves to increase market performance in liberalized electricity markets: The case of Austria, Doctoral Thesis, TU Vienna, 2003, p. 95

[Stad05] Stadler, M., Palensky, P., Lorenz, B., Weihs, M, Roesener, C.: Projekt IRON - Integral Resource Optimization Networks and their techno-economic constraints, International Journal of Distributed Energy Resources, Volume 1 Number 4, December 2005, pp. 299–320

[Stad06] Stadler, M., Firestone, R. M., Curtil, D., and Marnay, C.: On-Site Generation Simulation with EnergyPlus for Commercial Buildings, ACEEE Summer Study on Energy Efficiency in Buildings, August 13-18, 2006, Pacific Grove, California, ISBN 0-918249-56-2

[Stad07] Stadler, M., Kupzog, F., Palensky, P.: Distributed Energy Resource Allocation and Dispatch: an Economic and Technological Perception, International Journal of Electronic Business Management, Vol. 5, No. 3, 2007, pp. 182–196

[Stal05] Stallings, W.: Operating Systems, Internals and Design Principles, Pearson, Prentice Hall, 2005, pp. 399–411

[Stein05] Steinmetz, R., Wehrle, K.: Peer-to-Peer Systems and Applications, Springer, 2005, pp. 49–52

[Tan06] Tanenbaum, A. S., Steen, M. van, Distributed Systems: Principles and Paradigms, Pearson Prentice Hall, 2006, ISBN 0132392275

[Trey04] Treytl, A., Sauter, T., Bumiller, G.: Real-time Energy Management over Power-lines and Internet, International Symposium of Power-Line Communication and its Applications (ISPLC 2004), Zaragoza, Spain, April 2004.

[Trud06] Trudnowsky, D., Donelly, M., Lightner, E.: Power-System Frequency and Stability Control using Decentralized Intelligent Loads, Proceedings of IEEE PES T&D Conference and Exposition, Dallas, Texas, May 2006

[UCTE04a] UCTE Operation Handbook, Policy 1: Load-Frequency-Control and Performance, UCTE Brussels, Belgium 2004

References

[UCTE04b] UCTE Operation Handbook, Appendix 1: Load-Frequency-Control and Performance, UCTE Brussels, Belgium 2004

[UCTE05] UCTE Statistical Yearbook 2005, UCTE Brussels, Belgium 2005, pp. 106–107

[Vyv03] Vyver, J. v. d., Deconinck, G., Belmans, G.: The Need for a Distributed Algorithm for Control of the Electrical Power Infrastructure, Proc. of the International Symposium on Computational Intelligence for Measurement Systems and Applications, Lugarno, Switzerland, July 2003

[Wer01] Werner, T.G., Wirth, W., Vierheilig, N., Beissler, G.: Energy balance management systems for distribution companies, IEEE Porto Power Tech Proceedings, Porto, Sept. 2001

[Wol08] Wolfig, R.: A Distributed Platform for Integrated Modular Avionics, Doctoral Thesis, TU Vienna, June 2008

[Xu07] Xu, Z.;, Ostergaard, J., Togeby, M., Marcus-Moller, C.: Design and Modelling of Thermostatically Controlled Loads as Frequency Controlled Reserve, IEEE Power Engineering Society General Meeting, June 2007

[Yau90] Yau, T. S., Huff, R. G., Willis, H. L.: Demand-side management impact on the transmission and distribution system, IEE Transactions on Power Systems, Vol. 5, No. 2, May 1990

[Ygge96] Ygge, F, Gustavsson, R., Akkermans, H.: HOMEBOTS: Intelligent Agents for Decentralized Load Management, in proc. of Distribution Automation & Demand Side Management, Vienna, Austria, October 1996

Internet references

[1] "Energie der Zukunft", Funding scheme of the Austrian Federal Ministry of Innovation, Traffic and Technology and the Austrian Federal Ministry of Economy, *http://www.energiederzukunft.at*, visited 11/2007

[2] Energy call of the 7th Framework Program (FP7) of the European Union, *http://cordis.europa.eu/fp7/dc/index.cfm?fuseaction=UserSite.CooperationDetailsCallPage&call_id=80*, visited 11/2007

[3] "NEDO" (Department of the New Energy and Industrial Technology Development Organization, Japan), *http://www.nedo.go.jp/english*, visited 11/2007

[4] "GridWise", collaborative venture of the US Department of Energy and the GridWise Alliance, *http://www.gridwise.org*, visited 02/2008

[5] "Kops II" hydro storage and power plant *http://www.kopswerk2.at*, visited 11/2007

[6] "UCTE - Union for the Co-ordination of Transmission of Electricity", *http://www.ucte.org*, visited 02/2008

[7] "Verbund Netz APG", amount of primary control reserve in Austria 2007, *http://www.verbund.at/cps/rde/xchg/internet/hs.xsl/197_5436.htm*, visited 03/2008

[8] "MySQL: the world's most popular open source database", *http://www.mysql.com*, visited 03/2008

[9] "SmartGrids: European Technology Platform, Vision and Strategy for Europe's Electricity Networks of the Future." *http://www.smartgrids.eu*, visited 04/2008

[10] "RAKON SMD Microprocessor Crystal, Low cost resin sealed surface mount microprocessor crystal with max height 1.2mm", *http://www.rakon.com/models/display_model?model_id=131&action=display&reset=1*, visited 06/20

Südwestdeutscher Verlag für Hochschulschriften

Wissenschaftlicher Buchverlag bietet kostenfreie **Publikation** von **Dissertationen und Habilitationen**

Sie verfügen über eine wissenschaftliche Abschlußarbeit zu aktuellen oder zeitlosen Fragestellungen, die hohen inhaltlichen und formalen Anspruchen genügt, und haben **Interesse an einer honorarvergüteten Publikation?**

Dann senden Sie bitte erste Informationen über Ihre Arbeit per Email an: info@svh-verlag.de.

Unser Außenlektorat meldet sich umgehend bei Ihnen.

Südwestdeutscher Verlag für Hochschulschriften
Aktiengesellschaft & Co. KG

Dudweiler Landstr. 99
D – 66123 Saarbrücken
www.svh-verlag.de

Printed by Books on Demand GmbH, Norderstedt / Germany